TEXT BOOK
OF
CORRELATION AND REGRESSION

DPH MATHEMATICS SERIES

TEXT BOOK
OF
CORRELATION AND REGRESSION

By

A.K. Sharma

DISCOVERY PUBLISHING HOUSE
NEW DELHI-110002

First Published - 2005

Reprinted - 2017

ISBN: 978-81-7141-935-7

Text Book of Correlation and Regression

Published by:

DISCOVERY PUBLISHING HOUSE PVT. LTD.
4383/4B, Ansari Road, Darya Ganj
New Delhi-110 002 (India)
Phone: +91-11-23279245, 43596064-65
Fax: +91-11-23253475
E-mail: discoverypublishinghouse@gmail.com
sales@discoverypublishinggroup.com
web: www.discoverypublishinggroup.com

Printed at:
Infinity Imaging Systems
Delhi

Preface

This book on Text book of Correlation and Regression has been written to meet the requirement of graduate students of all Indian Universities. This book is also helpful for the preparation of I.A.S/ P.C.S and other competitions.

The subject matter has been presented in such a way that it is easily accessible to students. The proof of various theorems and examples have been given with minute details. Each chapter of this book contains complete theory and large number of solved examples we have been selected sufficient problems from various universities examination paper.

The author general home that the present book will be warmly received by the students and teachers. We shall indeed be very thankful to our colleagues for their recommend this book for their students.

A.K. Sharma

Contents

1

Correlation Analysis

INTRODUCTION

So far we have with the variable *e.g.*, the distribution of people according to heights or weights, annual rainfall of a certain area. The agricultural yield and so on. If two quantities vary in such a way that movements in one are accompanied by movements in the other, these quantities are correlated. For example, there exists some relationship between price of commodity and amount demanded, increase in rainfall up to a point and production of rice, an increase in the number of television licences and number of cinemagoers, etc. The degree of relationship between the variables under consideration is measured through the correlation analysis. The measure of correlation called the correlation coefficient or correlation index summarizes in one figure the direction and degree of correlation. The correlation analysis refers to the techniques used in measuring the closeness of the relationship between the variables.

The important definitions of correlation are given as follows :

1. "When the relationship is of a quantitative nature, the appropriate statistical tool for discovering and measuring the relationship and expressing it in self formula is known as correlation."

   —*Croxton & Cowden*

2. "Correlation analysis attempts to determine the '*degree of relationship*' between variables." —*Ya Lun Chou*

3. "Correlation is an analysis of the covariation between two or more variables." —*A.M. Tuttle*

4. "Correlation analysis deals with the association between two or more variables," —*Simpson & Kafka*

5. "If two or more quantities vary in sympathy so that movements in one and to be accompanied by corresponding movements in the other(s) then they are said to be correlated." —*L.R. Conner*

The problem of analysing the relation between different series should be broken into three steps :

1. Determining whether a relation exists and, if it does, measuring it.
2. Testing whether it is significant.
3. Establishing the cause and effect relation, if any.

An extremely high and significant correlation between the increase in smoking and increase in lung cancer would not prove that smoking causes lung cancer. The proof of a cause and effect relation can be developed only by means of an exhaustive study of the operative elements themselves.

It should be noted that the detection and analysis of correlation (*i.e.*, covariation) between two statistical variables requires relationship of some sort which associates the observation in pairs, one of each pair being a value of each of the two variables. In general, the pairing relationship may be of almost any nature, such a observations at the time or place or over a period of time of different places.

The computation concerning the degree of closeness is based on the regression equation. However, it is possible to perform correlation analysis without actually having a regression equation.

SIGNIFICANCE OF THE STUDY OF CORRELATION

The study of correlation is of immense use in practical life because of the following reasons :

(a) Once we know that two variables are closely related, we can estimate the value of one variable given the value of another. This is known with the help of regression analysis discussed in the next chapter.

(b) Correlation analysis contributes to the understanding of economic behaviour, aids in locating the critically important variables on which other depend, may reveal to the economist the connection by which disturbances spread and suggest to him the paths through which stabilizing forces may become effective.

(c) Most of the variables show some kind of relationship. For example, there is relationship between price and supply, income and expenditure, etc. With the help of correlation analysis we can measure in one figure the degree of relationship existing between the variables.

The correlation analysis enables the executive to estimate costs, sales, prices and other variables on the basis of some other series with which these costs, sales, or prices may be functionally related.

However, it should be noted that coefficient of correlation is one of the most widely used and also one of the most widely abused statistical measure. It is abused in the sense that one sometimes overlooks the fact that correlation measures are nothing but the strength of linear relationship and that it does not necessarily imply a cause-effect relationship.

(a) The effect of correlation is to reduce the range of uncertainly. The prediction based on correlation analysis is likely to be more valuable and near to reality.

(b) Progressive development in the methods of science and philosophy has been characterized by increase in the knowledge of relationship or correlations. In nature also one finds multiplicity of interrelated forces.

CORRELATION AND CAUSATION

Correlation analysis helps us in determining the degree of relationship between two or more variables. The explanation of a significant degree of correlation may be any one. Combination of the following reasons :

(i) Both the variables may be mutually influencing each other so that neither can be designated as the cause and the other the effect. There may be a high degree of correlation between the variables but it may be difficult to pinpoint as to which is the cause and which is the effect. This is especially likely to be so in case of economic variables. For example, such variables as demand and supply, price and production, etc. mutually interact. To take a specific case, it is a well-known principle of economics that as the price of a commodity increases its demand goes down and so price is the cause and demand the effect. But it is also possible that increased demand of a commodity due to growth of population or other reasons may exercise an upward pressure on price. Now, the cause is the increased demand, the effect the price. Thus, at times it may become difficult to explain from the two correlated variables which is the cause and which is the effect because both may be resting on each other.

(ii) Both the correlated variables may be influenced by one or more other variables. It is just possible that a high degree of correlation between the variables may be due to some causes affecting each variable or different causes affecting each with the same effect. For example, a high degree of correlation between the yield per acre of rice and tea may be due to the fact that both are related to the amount of rainfall. But none of the two variables is the cause of the other. To take another example : suppose the correlation of teachers salaries and the consumption of liquor over a period of year comes out to be 0.9, this does not prove that teachers drink : nor does it prove that liquor sale increases teachers salaries. Instead, both variables move together because both are influenced by a third variable—Long-run growth in national income and population.

(iii) The correlation may be due to pure chance, especially in a small sample. We may get a high degree of correlation between two variables in a sample but in the universe there may not be any relationship between the variables at all. This is especially so in case of small samples. Such a correlation may arise either because of pure random sampling variation or because of the bias of the investigator in selecting the sample. The following example shall illustrate the point:

Income (Rs.) :	3500	3600	3700	3800	3900
Weight (lbs) :	120	140	160	180	200

The above data show a perfect positive relationship between income and weight, *i.e.*, as the income is increasing the weight is increasing and the rate of change between two variables is the same.

The above points clearly bring out the fact that a mathematical relationship implies nothing in itself about cause and effect. In many instances extremely high degree of correlation between two variables may be obtained when no meaning can be attached to the answer. There is, for example, extremely high correlation between some series representing the production of pigs and the production of pig iron, yet no one has ever believed that this correlation has any meaning or that it indicates the existence of a cause-effect relation. By itself, it establishes only covariation. Correlation observed between variables that cannot conceivably be casually related is called spurious or nonsense correlation. More appropriately, we should remember that it is interpretation of the degree of correlation that is spurious, not the degree of correlation itself. The high degree of correlation indicates only the mathematical result. We should reach a conclusion based on logical reasoning and intelligent investigation of significantly related matters. It may also be pointed out that errors in correlation analysis include not only reading causation into spurious correlation but also interpreting spuriously a perfectly valid relationship.

TYPES OF CORRELATION

Correlation classified in several different ways. The most important correlation are as follows :

1. Positive or negative.
2. Simple, partial and multiple.
3. Linear and non-linear.

Positive and Negative Correlation

If one variable is increasing the other is also increasing or if one variable is decreasing other is also decreasing the correlation is said to be positive. If on the other hand the variables are varying in opposite direction the correlation is said to be negative.

I. Positive Correlation

X :	10	12	15	18	20	X :	80	70	60	40	30
Y :	15	20	22	25	37	Y :	50	44	30	20	10

II. Negative Correlation

X :	20	30	40	60	80	X :	100	90	60	40	30
Y :	40	30	22	15	10	Y :	10	20	30	40	50

Multiple Correlation

The term *multiple correlation* refers to the theory of correlation involving more than two varaibles. We have seen in the previous chapter that simple correlation deals with the degree of relationship between two variables, such as ages of hwbands and wives; supply and demand of a commodity, height and weight, and so on. (Multiple correlation is used to find the degree of inter-relationship among threc or more variables). For example the yield of crop in a year may depend upon rainfall, manure, the average temperature and average humidity during the period between sowing and harvesting of the crop; the results of houses may depend upon tax rates as well as upon building costs and upon other variable also; general inte illigence in schools may be related to grades in mathematics and grades in English and so on. Thus the aim of the theory of multiple correlation is to know how far the *dependent variable* is influenced by the *independent variables*. We shall denote the multiple correlation between x_i dependent variable and independent variables, by $R_{1.234...n}$.

Partial Correlation

The simple correlation between two variates when the influence of other variates in them has been eliminated from both is called *partial correlation*. Thus we may measure the correlation between the heights and weights of boys of the same age, say 15 years, fiwi the age factor is the same for all the boys and weights and heights are "variable factors. Again we may measure the correlations between the age at marriage and the number of children in families of the same income group. Here the factor income per family is constant. The correlation between grades in mathematics and gradestn English for higher secondary boys of Delhi schools having the same inteligence quotients of 90, is also an example of partial correlation. A classical example is the correlation between statures of father an sons, when the stature of the mother has a particular value, say 60 inches.

Linear and Non-Linear (Curvilinear) Correlation

The distinction between linear and non-linear correlation is based upon the constancy of the ratio of change between the variables. If the amount of change in one variable tends to bear constant ratio to the amount of change in the other variable then the correlation is said to be linear. For example, observe the following two variables X and Y :

X :	20	40	60	80	100
Y :	30	60	90	120	150

It is clear that the ratio of change between the two variables is the same. If such variables are plotted on a graph paper all the plotted points would fall on a straight line.

Correlation would be called not-linear or curvilinear if the amount of change in one variable does not bear a constant ratio to the amount of change in the other variable.

The following two diagrams will illustrate the difference between linear and curvilinear correlation :

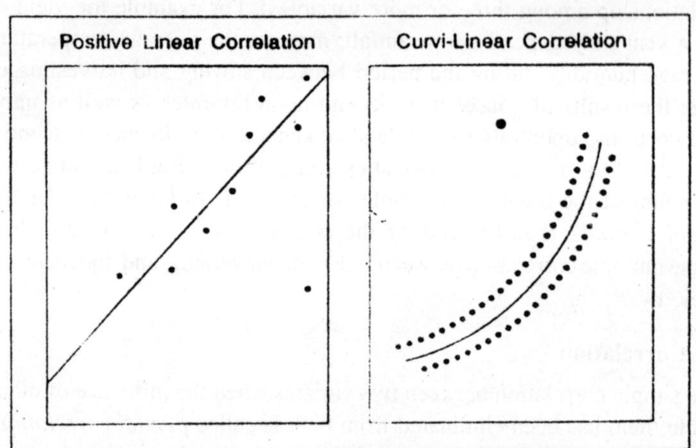

METHODS OF STUDYING CORRELATION

The correlation may be studying by following ways:

(a) Concurrent Deviation Method
(b) Karl Pearson's Coefficient of Correlation
(c) Graphic Method
(d) Scatter Diagram Method

(e) Method of Least Squares.

Of these, the first two are based on the knowledge of diagrams and graphs, whereas the others are the mathematical methods. Each of these methods shall be discussed in detail in the following pages.

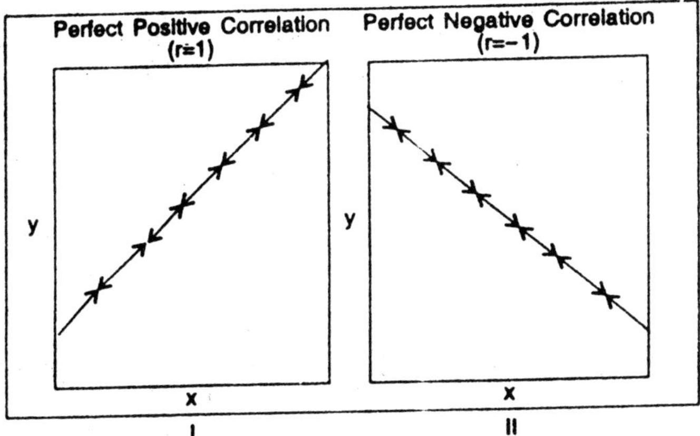

Scatter Diagram Method

The simplest device for ascertaining whether two variables are related is to prepare a dot chart called scatter diagram. When this method is used the given data are plotted on a graph paper in the form of dots, *i.e.*, for each pair of X and Y values we put a dot and thus obtain as many points as the

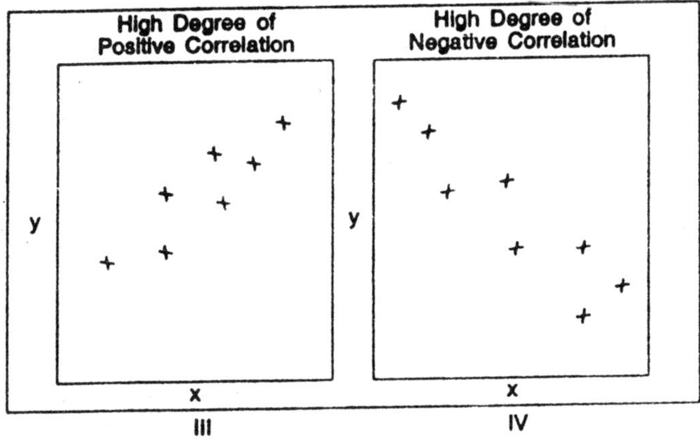

number of observations. The more closely the points come to a straight line, failing from the lower left-hand corner to the upper right-hand corner, correlation is said to be perfectly positive (*i.e.* r = + 1) (diagram I). On the other hand, if all the points are lying on a straight line rising from the upper left-hand corner to the lower right-hand corner of the diagram, correlation is said to be perfectly negative (*i.e.* r = – 1) (diagram II). If the plotted points fall in a narrow band there would be a high degree of correlation between the variables—correlation shall be positive, if the points show a rising tendency from the lower left-hand corner to the upper right-hand corner (diagram III) and negative if the points show a declining tendency from the upper left-hand corner to the lower right-hand corner of the diagram (diagram IV). On the other hand, if the points are widely scattered over the diagram it indicates very little relationship between the variables—correlation shall be positive if the points are rising from the lower left-hand corner to the upper right-

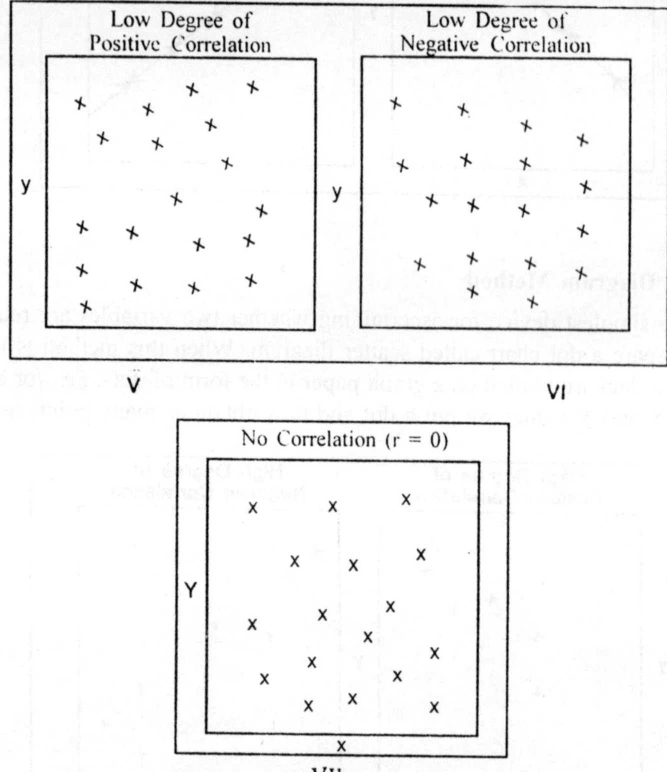

hand corner (diagram V) and negative if the points are running from the upper left-hand side to the lower right-hand side of the diagram (diagram VI). If the plotted points lie on a straight line parallel to the X-axis or in a haphazard manner, it shows absence of any relationship between the variables (*i.e.*, r = 0) as shown in diagram VII.

Example:

Given the following pairs of values of the variables X and Y :

X :	*2*	*3*	*5*	*6*	*8*	*9*
Y :	*6*	*5*	*7*	*8*	*12*	*11*

(a) *Make a scatter diagram.*

(b) *Is there any correlation between the variables X and Y?*

(c) *By graphic inspection, draw an estimating line.*

Solution:

By looking at the scatter diagram we can say that the variables X and Y are correlated. Further, correlation is positive because the trend of the points is upward rising from the lower left-hand corner to the upper right-hand corner of the diagram. The diagram also indicates that the degree of relationship is higher because the plotted points are near to the line which shows perfect relationship between the variables.

MERITS AND LIMITATIONS OF THE METHOD

Merits

Following are the merits of scatter diagram method :

(a) It is simple and non-mathematical method of studying correlation between the variables. As such it can be easily understood and a rough idea can very quickly be formed as to whether or not the variables are related.

(b) It is not influenced by the size of extreme items whereas most of the mathematical methods of finding correlation are influenced by extreme items.

(c) Making a scatter diagram usually is the first step in investigating the relationship between two variables.

Limitations

By applying this method we can get an idea about the direction of correlation and also whether it is high or low. But we cannot establish the exact degree of correlation between the variables as is possible by applying the mathematical methods.

Graphic Method

When this method is used the individual values of the two variables are plotted on the graph paper. We thus obtain two curves, one for X variable and another for Y variable. By examining the direction and closeness of the two curves so drawn we can infer whether or not the variables are related.

If both the curves drawn on the graph are moving in the same direction (either upward or downward) correlation is said to be positive. On the other hand, if the curves are moving in the opposite directions correlation is said to be negative.

Example:

From the following data ascertain whether the income and expenditure of the 100 workers of a factory are correlated :

Year	Average income (in Rs.)	Average expenditure (in Rs.)	Year	Average income (in Rs.)	Average expenditure (in Rs.)
1989	3100	3000	1994	4300	4100
1990	3320	3200	1995	4500	4200
1991	3500	3350	1996	4650	4400
1992	3700	3500	1997	4800	4650
1993	4200	4000	1998	5000	4500

Solution:

The following graph shows that the variables, income and expenditure, are closely related.

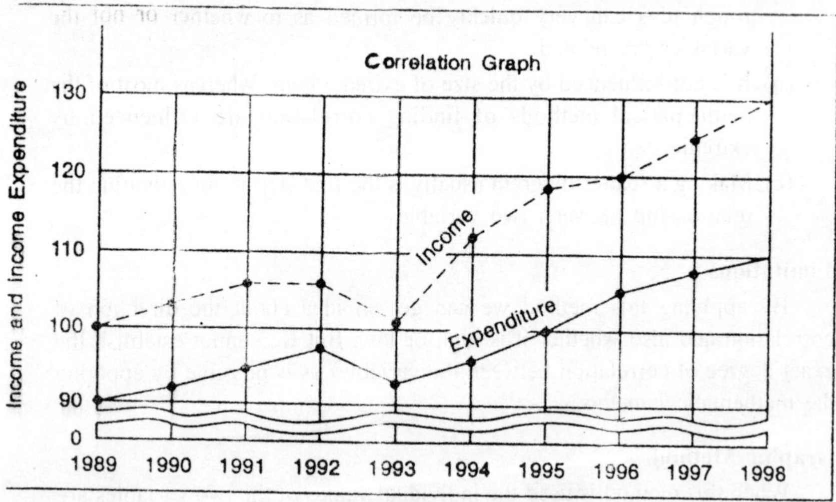

KARL PEARSON'S COEFFICIENT OF CORRELATION

The Pearson coefficient of correlation is denoted by the symbol r. It is one of the very few symbols that are used universally for describing the degree of correlation between two series. The formula for computing Pearsonian r is :

$$r = \frac{\Sigma xy}{N \sigma_x \sigma_y} \quad \text{...(i)}$$

Here $x = (X - \overline{X}); \; y = (Y - \overline{Y})$

σ_x = Standard deviation of series X

σ_y = Standard deviation of series Y

N = Number of pairs of observations

r = the (product moment) correlation coefficient.

The value of the coefficient of correlation as obtained by the above formula shall always lie between ± 1. When r = + 1, it means there is perfect positive correlation between the variables. When r = – 1, it means there is perfect negative correlation between the variables. When r = 0, it means there is no relationship between the two variables. However, in practice such values of r as + 1, – 1, and 0 are rare. We normally get values which lie between + 1 and – 1 such as + 0.6, – 0.46, etc. The coefficient of correlation describes not only the magnitude of correlation but also its direction. Thus, + 0.6 would mean that correlation is positive because the sign of r is + and the magnitude of correlation is 0.6. Similarly – 0.46 means low degree of negative correlation.

The above formula for computing Pearson's coefficient of correlation can also be written as:

$$r^* = \frac{\Sigma xy}{\sqrt{\Sigma x^2 \times \Sigma y^2}} \quad \text{...(ii)}$$

where $x = (X - \overline{X})$ and $y = (Y - \overline{Y})$

It is obvious that while applying this formula we have not to calculate separately the standard deviation of X and Y series as is required by formula (i). This simplifies greatly the task of calculating correlation coefficient.

Steps.

(a) Take the deviations of Y series from the mean of Y and denote these deviations by y.

(b) Square these deviations and obtain the total, *i.e.*, Σx^2.

(c) Multiply the deviations of X and Y series and obtain the total, *i.e.*, Σxy.

(d) Substitute the values of Σxy, Σx^2 and Σy^2 in the above formula.

(e) Take the deviations of X series from the mean of X and denote these deviations by x.

(f) Square these deviations and obtain the total, *i.e.*, Σx^2.

The following examples will the procedure :

Example 1:

The following able gives indices of industrial production of registered unemployed (in hundred thousand). Calculate the value of the coefficient so obtained.

Year	:	*1991*	*1992*	*1993*	*1994*	*1995*	*1996*	*1997*	*1998*
Index of Production	:	*100*	*102*	*104*	*107*	*105*	*112*	*103*	*99*
Number Unemployed	:	*15*	*12*	*13*	*11*	*12*	*12*	*19*	*26*

Solution:

Calculation of Karl Pearson's Correlation Coefficients

Year	**Production X**	**$(X - \overline{X})$ x**	**x^2**	**Unemployed Y**	**$(Y - \overline{Y})$ y**	**y^2**	**xy**
1991	100	− 4	16	15	0	0	0
1992	102	− 2	4	12	− 3	9	+ 6
1993	104	0	0	13	− 2	4	0
1994	107	+ 3	9	11	− 4	16	− 12
1995	105	+ 1	1	12	− 3	9	− 3
1996	112	+ 8	64	12	− 3	9	− 24
1997	103	− 1	1	19	+ 4	16	− 4
1998	99	− 5	25	26	+ 11	121	− 55
	$\Sigma X - 832$	$\Sigma x = 0$	$\Sigma x^2 = 120$	$\Sigma Y = 120$	$\Sigma y = 0$	$\Sigma y^2 = 184$	$\Sigma xy = -92$

$$r = \frac{\Sigma xy}{\sqrt{\Sigma x^2 . \Sigma y^2}}$$

$$x = (X - \overline{X});\ y = (Y - \overline{Y})$$

$$\overline{X} = \frac{\Sigma X}{N} = \frac{832}{8} = 104;$$

$$\overline{Y} = \frac{\Sigma Y}{N} = \frac{120}{8} = 15$$

$$\Sigma xy = 92,\ \Sigma x^2 = 120,\ \Sigma y^2 = 184$$

$$r = \frac{-92}{\sqrt{120 \times 184}} = -0.619$$

Example 2:

Making use of the data summarized below, calculate the coefficient of correlation, r_{12} :

Case	X_1	X_2	*Case*	X_1	X_2
A	*10*	*9*	*E*	*12*	*11*
B	*6*	*4*	*F*	*13*	*13*
C	*9*	*6*	*G*	*11*	*8*
D	*10*	*9*	*H*	*9*	*4*

Solution:

Calculation of Coefficient of Correlation

Case	X_1	$(X_1 - \overline{X}_1)$ x_1	x_1^2	X_2	$(X_2 - \overline{X}_2)$ x_2	x_2^2	$x_1 x_2$
A	10	0	0	9	+ 1	1	0
B	6	– 4	16	4	– 4	16	16
C	9	– 1	1	6	– 2	4	2
D	10	0	0	9	+ 1	1	0
E	12	+ 2	4	11	+ 3	9	6
F	13	+ 3	9	13	+ 5	25	15
G	11	+ 1	1	8	0	0	0
H	9	– 1	1	4	– 4	16	4
N = 8	$\Sigma X_1 = 80$	$\Sigma x_1 = 0$	$\Sigma x_1^2 = 32$	$\Sigma X_2 = 64$	$\Sigma x_2 = 0$	$\Sigma x_2^2 = 72$	$\Sigma x_1 x_2 = 43$

$$\overline{X}_1 = \frac{\Sigma X_1}{N} = \frac{80}{8} = 10$$

$$\overline{X}_2 = \frac{\Sigma X_2}{N} = \frac{64}{8} = 8$$

$$r_{12} = \frac{\Sigma x_1 x_2}{\sqrt{\Sigma x_1^2 \times \Sigma x_2^2}}$$

$$\Sigma x_1 x_2 = 43, \Sigma x_1^2 = 32, \Sigma x_2^2 = 72$$

Substituting the value

$$r_{12} = \frac{43}{\sqrt{32 \times 72}} = \frac{43}{\sqrt{2304}} = \frac{43}{48} = +0.896$$

Direct Method of Finding out Correlation Coefficient

Correlation coefficient can also be calculated without taking deviations of items either from actual mean or assumed mean, *i.e.*, actual X and Y values. The formula in such a case is :

$$r = \frac{N\Sigma XY - (\Sigma X)(\Sigma Y)}{\sqrt{N\Sigma X^2 - (\Sigma X)^2}\sqrt{N\Sigma Y^2 - (\Sigma Y)^2}}$$

This formula would give the same answer as we get when deviations of items are taken from actual mean or assumed mean.

Example 1:

Calculate correlation coefficient from data of illustration 3 by the direct method, i.e., without taking the deviation of items from actual or assumed mean.

Solution:

We have that

$$r = \frac{N\Sigma XY - (\Sigma X)(\Sigma Y)}{\sqrt{N\Sigma X^2 - (\Sigma X)^2}\sqrt{N\Sigma Y^2 - (\Sigma Y)^2}}$$

Calculation of Correlation Coefficient (Direct Method)

X	X^2	Y	Y^2	XY
9	81	15	225	135
8	64	16	256	128
7	49	14	196	98
6	36	13	169	78
5	25	11	121	55
4	16	12	144	48
3	9	10	100	30
2	4	8	64	16
1	1	9	81	9
$\Sigma X = 45$	$\Sigma X^2 = 285$	$\Sigma Y = 108$	$\Sigma Y^2 = 1{,}356$	$\Sigma XY = 597$

$N = 9, \Sigma XY = 597,$

$\Sigma X = 45, \Sigma Y = 108,$

$\Sigma X^2 = 285, \Sigma Y^2 = 1{,}356$

$$r = \frac{9 \times 597 - 45 \times 108}{\sqrt{9 \times 285 - (45)^2}\ \sqrt{9 \times 1356 - (108)^2}}$$

$$r = \frac{5373 - 4860}{\sqrt{2565 - 2025}\ \sqrt{12{,}204 - 11{,}664}}$$

$$r = \frac{513}{\sqrt{540 \times 540}} = \frac{513}{540} = +0.95.$$

Example 2:

Calculate coefficient of correlation from the following data :

X :	*100*	*200*	*300*	*400*	*500*	*600*	*700*
Y :	*30*	*50*	*60*	*80*	*100*	*110*	*130*

Solution:

To simplify calculation let every value of X be divided by 100 and every value of Y by 10 and denote these series by X' and Y'.

Computation of Correlation Coefficient

X	X/100 X'	$(X' - \overline{X})$ x	x^2	Y x^2	Y/10 Y'	$(Y' - \overline{Y})$ y	y^2	xy
100	1	– 3	9	30	3	– 5	25	15
200	2	– 2	4	50	5	– 3	9	6
300	3	– 1	1	60	6	– 2	4	2
400	4	0	0	80	8	0	0	0
500	5	1	1	100	10	+ 2	4	2
600	6	+ 2	4	110	11	+ 3	9	6
700	7	+ 3	9	130	13	+ 5	25	15
		$\Sigma x = 0$	$\Sigma x^2 = 28$			$\Sigma y = 0$	$\Sigma y^2 = 76$	$\Sigma xy = 46$

$$r = \frac{\Sigma xy}{\sqrt{\Sigma x^2 . \Sigma y^2}}$$

$$x = (X - \overline{X});\ y = (Y - \overline{Y})$$

$$\Sigma\, xy = 46,\ \Sigma\, x^2 = 28,\ \Sigma\, y^2 = 76$$

$$r = \frac{46}{\sqrt{28 \times 76}} = +0.997$$

When Deviations are taken from an Assumed Mean

When deviations are taken from an assumed mean the following formula is applicable :

$$r = \frac{N\Sigma d_x d_y - \Sigma d_x \times \Sigma d_y}{\sqrt{N\Sigma d_x^2 - (\Sigma d_x)^2}\sqrt{N\Sigma d_y^2 - (\Sigma d_y)^2}}$$

where d_x refers to deviations of X series from an assumed mean, *i.e.*, $(X - \overline{X})$.

Similarly, d_y refers to deviations of Y series from an assumed mean *i.e.*, $(Y - \overline{Y})$.

$\Sigma d_x d_y$ = sum of the product of the deviations of X and Y series from their assumed means.

Σd_x^2 = sum of the squares of the deviations of X series from an assumed means.

Σd_y^2 = sum of the squares of the deviations of Y series from an assumed means.

Σd_x = sum of the deviations of X series from an assumed means.

Σd_y = sum of the deviations of Y series from an assumed means.

It may be pointed out that there are many variations of the above formula. For example, the above formula may be written as :

$$r = \frac{N\Sigma d_x d_y - \{(\Sigma d_x) \times (\Sigma d_y)\}}{\sqrt{N\Sigma d_x^2 - (\Sigma d_x)^2}\sqrt{N\Sigma d_y^2 - (\Sigma d_y)^2}}$$

But the formula given above is the easiest to apply.

Steps :

(a) Substitute the value of $\Sigma d_x d_y$, Σd_x . Σd_y, Σd_x^2 and Σd_y^2 in the formula given above.

(b) Multiply d_x and d_y and obtain the total $\Sigma d_x d_y$.

(c) Square d_y and obtain the total Σd_y^2.

(d) Square d_x and obtain the total Σd_x^2.

(e) Take the deviations of Y series from an assumed mean and denote these deviations by d_y and obtain the total *i.e.* Σd_y.

(f) Take the deviations of X series from an assumed mean and denote these deviations by d_x and obtain the total, *i.e.*, Σd_x.

It may be pointed out that this form is simplest to apply. The following examples shall illustrate the procedure.

Example 3:

Given :

Total of the product of deviations of X and Y series = 3,044

Number of pairs of observations = 10

Total of the deviations of X series = – 170

Total of the deviations of Y series = – 20

Total of the squares of deviations of X series = 8,288

Total of the squares of deviations of Y series = 2,264.

Find out the coefficient of correlation when the assumed means of X series and Y series are 82 and 68 respectively.

Solution:

We are given :

$\Sigma d_x d_y = 3{,}044,\ \Sigma d_x = -170,\ \Sigma d_y = -20,$

$\Sigma d_x^2 = 8{,}288,\ \Sigma d_y^2 = 2{,}264,\ N = 10.$

Applying the formula $r = \dfrac{N \Sigma d_x d_y - \Sigma d_x \Sigma d_y}{\sqrt{N \Sigma d_x^2 - (\Sigma d_x)^2}\ \sqrt{N \Sigma d_y^2 - (\Sigma d_y)^2}}$

$$= \frac{10 \times 3044 - (-170)(-20)}{\sqrt{10 \times 8288 - (-170)^2}\ \sqrt{10 \times 2264 - (20)^2}}$$

$$= \frac{30444 - 3400}{\sqrt{82880 - 28900}\ \sqrt{22640 - 400}}$$

$$= \frac{27040}{\sqrt{53980}\ \sqrt{22240}}$$

$$= \frac{27040}{232.336 \times 149.131} = \frac{27040}{34648.5} = +0.7804$$

Another form of the formula when deviations from the assumed average.

$$r = \frac{\Sigma d_x d_y - N(\overline{X} - A_x)(\overline{Y} - A_y)}{N \sigma_x \sigma_y}$$

Where $\Sigma d_x d_y$ = sum of deviations from the assumed average.

$\overline{X}$ = actual mean of the X series,

$\overline{Y}$ = actual mean of Y series,

A_x = assumed mean of X series,

A_y = assumed mean of Y series.

The following example shall illustrate the application of the above formula.

Example 4:

Given :

Number of pairs of observation X and Y series = 8

X series Arithmetic Average = 74.5

X series Assumed Average = 69.0

X series Standard Deviation = 13.07

Y series Arithmetic Average = 125.5

Y series Assumed Average = 112

Y series Standard Deviation = 15.85

Summation of products of corresponding deviations of X and Y series = 2,176. Calculate the coefficient of correlation.

Solution:

$\Sigma d_x d_y$ = sum of the products of deviations from the assumed average.

$\overline{X}$ = actual mean of the X series.

$\overline{Y}$ = actual mean of the Y series.

A_x = actual mean of the X series.

A_y = actual mean of the Y series.

We are given : $\Sigma d_x d_y = 2{,}176$, $\overline{X} = 74.5$,

$A_x = 69$, $\sigma_x = 13.07$

$\overline{Y} = 125.5$, $A_y = 112$,

$N = 8$, $\sigma_y = 15.85$

Substituting the values in the formula

$$r = \frac{2176 - 8(74.5 - 69)\,(125.5 - 112)}{8 \times 13.07 \times 15.85}$$

$$= \frac{2176 - 594}{1657.276} = \frac{1582}{1657.276} = +0.955$$

Example 5:

The following table gives the distribution of items of production and also the relatively defective items among them, according to size groups. Find the correlation coefficient between size and defect in quality and its probable error.

Size-group :	*15–16*	*16–17*	*17–18*	*18–19*	*19–20*	*20–21*
No. of items :	*200*	*270*	*340*	*360*	*400*	*300*
No. of defective items :	*150*	*162*	*170*	*180*	*180*	*114*

Solution:

Let mid-point of size be denoted by X and % of defective items by Y.

Calculation of Correlation Coefficient

Size	m.p. X	(X – 17.5) d_x	d_x^2	% of Defective items Y	d_y	d_y^2	$d_x d_y$
15–16	15.5	– 2	4	75	+ 25	625	– 50
16–17	16.5	– 1	1	60	+ 10	100	– 10
17–18	17.5	0	0	50	0	0	0
18–19	18.5	+ 1	1	50	0	0	0
19–20	19.5	+ 2	4	45	– 5	25	– 10
20–21	20.5	+ 3	9	38	– 12	144	– 36
		$\Sigma d_x = 3$	$\Sigma d_x^2 = 19$		$\Sigma d_y = 18$	$\Sigma d_y^2 = 894$	$\Sigma d_x d_y = -106$

$$r = \frac{N\Sigma d_x d_y - \Sigma d_x \Sigma d_y}{\sqrt{N\Sigma d_x^2 - (\Sigma d_x)^2}\sqrt{N\Sigma d_y^2 - (\Sigma d_y)^2}}$$

$$= \frac{6 \times -106 - 3 \times 18}{\sqrt{6 \times 19 - (3)^2}\sqrt{6 \times 894 - (18)^2}}$$

$$= \frac{-636 - 54}{\sqrt{105}\sqrt{5040}} = -\frac{690}{727.46} = -0.949$$

$$P.E._r = 0.6745 \frac{1 - r^2}{\sqrt{N}} = 0.6745 \frac{1 - (8)^2}{\sqrt{6}}$$

$$= 0.6745 \times \frac{.36}{2.449} = 0.099$$

Example 6:

Calculate the coefficient of correlation between X and Y from the following data and calculate problem errors. Assume 69 and 112 as the mean value for X and Y respectively.

X :	*78*	*89*	*99*	*60*	*59*	*79*	*68*	*61*
Y :	*125*	*137*	*156*	*112*	*107*	*136*	*123*	*108*

Correlation of Grouped Data

The class intervals for Y are listed in the captions or column headings, and those for X are listed in the stubs at the left of the table, the order can also be reversed. The frequencies for each cell of the table are determined by either tallying or card sorting just as in the case of frequency distribution of a single variable.

The formula for calculating the coefficient of correlation is :

Steps.

a. Substitute the values of $\Sigma f d_x d_y$, $\Sigma f d_x$, $\Sigma f d_x^2$, $\Sigma f d_y$ and $\Sigma f d_y^2$ in the above formula and obtain the value of r.

b. Take the squares of the deviations of the variable Y and multiply them by the respective frequencies and obtain $\Sigma f d_y^2$.

c. Multiply the frequencies of the variable Y by the deviations of Y and obtain the total $\Sigma f d_y$.

d. Take the squares of the deviations of variable Y and multiply them by the respective frequencies and obtain $\Sigma f d_x^2$.

e. Multiply the frequencies of the variable X by the deviations of X and obtain the total $\Sigma f d_x$.

f. Add together all the cornered values as calculated in step (iii) and obtain the total $\Sigma f d_x d_y$.

g. Multiply $d_x d_y$ and respective frequency of each cell and write the figure obtained in right-hand upper corner of each cell.

h. Take the step deviations of the variable Y and denote these deviations by d_y.

i. Take the step deviations of the variable X and denote these deviations by d_x.

Example 1:

From the following data compute the coefficient of correlation between age of husbands and age of wives :

Age of husbands	*Age of wives*						*Total*
	15–25	***25–35***	***35–45***	***45–55***	***55–65***	***65–75***	
15–25	*1*	*1*					*2*
25–35	*2*	*12*	*1*				*15*
35–45	–	*4*	*10*	*1*			*15*
45–55		–	*3*	*6*	*1*		*10*
55–65			–	*2*	*4*	*2*	*8*
65–75				–	*1*	*2*	*3*
Total	*3*	*17*	*14*	*9*	*6*	*4*	*53*

Solution:

Computation of Correlation Coefficient

Y \ X		15–25	25–35	35–45	45–55	55–65	65–75				
Y	d_y \ d_x	– 2	– 1	0	+ 1	+ 2	+ 3	f	$f\,d_y$	$f\,d_y^2$	$f\,d_xd_y$
15–25	– 2	4 1	+ 2	–	–	–	–	2	– 4	8	6
25–35	– 1	+ 4 2	12 12	1	–	–	–	15	– 15	15	16
35–45	0	–	0 4	0 10	0 1	–	–	15	0	0	0
45–55	+ 1	–	–	0 3	6 6	2 1		10	+ 10	10	8
55–65	+ 2	–	–	–	4 2	16 4	+ 12 2	8	+ 16	32	32
65–75	+ 3	–	–	–	–	6 1	18 2	3	+ 9	27	24
f		3	17	14	9	6	4	N = 53	Σfd_y + 16	Σfd_y^2 = 92	Σfd_xd_y = 86
$f\,d_x$		– 6	– 17	0	9	12	12	Σfd_x = + 10			
$f\,d_x^2$		12	17	0	9	24	36	Σfd_x^2 = 98			
$f\,d_xd_y$		8	14	0	10	24	30	Σfd_xd_y = 86			

$$r = \frac{N\Sigma f d_x d_y - \Sigma f d_x \Sigma f d_y}{\sqrt{N\Sigma f d_x^2 - (\Sigma f d_x)^2}\sqrt{N\Sigma f d_y^2 - (\Sigma f d_y)^2}}$$

$$= \frac{53 \times 86 - 10 \times 16}{\sqrt{53 \times 98 - (10)^2}\sqrt{53 \times 92 - (16)^2}}$$

$$= \frac{4558 - 160}{\sqrt{5194 - 100}\sqrt{4876 - 256}}$$

$$= \frac{4398}{\sqrt{5094}\sqrt{4620}}$$

$$= \frac{4398}{4851.214} = 0.907$$

Example 2:

The following are the marks obtained by the students of a class in Statistics of Accountancy :

Roll No. of Students	***Marks in Statistics***	***Marks in Accountancy***	***Roll No. of Students***	***Marks in Statistics***	***Marks in Accountancy***
1	*15*	*13*	*13*	*14*	*11*
2	*0*	*1*	*14*	*9*	*3*
3	*1*	*2*	*15*	*8*	*5*
4	*3*	*7*	*16*	*13*	*4*
5	*16*	*8*	*17*	*10*	*10*
6	*2*	*9*	*18*	*13*	*11*
7	*18*	*12*	*19*	*11*	*14*
8	*5*	*9*	*20*	*11*	*7*
9	*4*	*17*	*21*	*12*	*18*
10	*17*	*16*	*22*	*18*	*15*
11	*6*	*6*	*23*	*9*	*15*
12	*19*	*18*	*24*	*7*	*3*

Prepare a correlation table taking the magnitude of each class interval as four marks and the list interval as equal to 0 and less than 4. Calculate Karl Pearson's coefficient of correlation between the marks in Statistics and marks in Accountancy and comment on the correlation table.

Solution:

Preparation of Correlation Table

Marks in Accountancy	Marks in Statistics 0–4	4–8	8–12	12–16	16–20	Total
0–4	| (2)	| (1)	| (1)			4
4–8	| (1)	| (1)	| (2)	| (1)		5
8–12	| (1)	| (1)	| (1)	| (2)	| (1)	6
12–16			| (2)	| (1)	| (2)	5
16–20		| (1)		| (1)	| (2)	4
Total	4	4	6	5	5	24

Let marks in Statistics be denoted by X and marks in Accountancy by Y.

Calculation of Correlation Coefficient

Y \ X	m.p.	m.p.	0–4 2	4–8 6	8–12 10	12–16 14	16–20 16				
Y m.p.		d_x / d_y	– 2	– 1	0	1	2	f	fd_y	fd_y^2	fd_xd_y
0–4	2	– 2	+8 2	+ 2 1	0 1			4	– 8	16	10
4–8	6	– 1	2 1	1 1	0 2	– 1 1		5	– 5	6	2
8–12	10	0	0 1	0 1	0 1	0 2	0 1	6	0	0	0
12–16	14	1			0 2	1 1	4 2	5	5	5	5
16–20	18	2		2 1		2 1	8 2	4	8	16	8
		f	4	4	6	5	5	N = 24	$\Sigma fd_y = 0$	$\Sigma fd_y^2 = 42$	$\Sigma fd_xd_y = 25$
		fd_x	– 8	– 4	0	5	10	$\Sigma fd_x = 3$			
		fd_x^2	16	4	0	5	20	$\Sigma fd_x^2 = 45$			
		fd_xd_y	10	1	0	2	12	$\Sigma fd_xd_y = 25$			

$$r = \frac{N\Sigma f d_x d_y - \Sigma f d_x \Sigma f d_y}{\sqrt{N\Sigma f d_x^2 - (\Sigma f d_x)^2}\sqrt{N\Sigma f d_y^2 - (\Sigma f d_y)^2}}$$

$$= \frac{24 \times 25 - 3 \times 0}{\sqrt{24 \times 45 - (3)^2}\sqrt{24 \times 42}} = \frac{600}{\sqrt{1071 \times 1008}} = \frac{600}{1039.02} = 0.578.$$

Example 3:

The following table gives the frequency, according to groups of marks obtained by 67 students in an intelligence test. Measure the degree of relationship between age and inteligence test:

Test marks	***Age in years*** 18	19	20	21	***Total***
200–250	*4*	*4*	*2*	*1*	*11*
250–300	*3*	*5*	*4*	*2*	*14*
300–350	*2*	*6*	*8*	*5*	*21*
350–400	*1*	*4*	*6*	*10*	*21*
Total	*10*	*19*	*20*	*18*	*69*

Solution:

Calculation of Coefficient of Correlation

X / Y		Age in years 18	19	20	21				
	d_y \ d_x	− 1	0	+ 1	+ 2	f	f d_y	f d_y^2	f $d_x d_y$
200–250	− 1	(+4) 4	(0) 4	(− 2) 2	(− 2) 1	11	− 11	11	10
250–300	0	(0) 3	(0) 5	(0) 4	(0) 2	14	0	0	0
300–350	+ 1	(− 2) 2	(0) 6	(8) 8	(10) 5	21	21	21	16
350–400	+ 2	(− 2) 1	(0) 4	(12) 6	(40) 10	21	42	84	50
Total		10	19	20	18	N = 67	Σfd_y = 52	Σfd_y^2 = 116	$\Sigma fd_x d_y$ = 66
f d_x		− 10	0	20	36	Σfd_x = 46			
f d_x^2		10	0	20	72	Σfd_x^2 = 102			
f $d_x d_y$		0	0	18	48	$\Sigma fd_x d_y$ = 66			

$$r = \frac{N\Sigma f d_x d_y - \Sigma f d_x \Sigma f d_y}{\sqrt{N\Sigma f d_x^2 - (\Sigma f d_x)^2}\sqrt{N\Sigma f d_y^2 - (\Sigma f d_y)^2}}$$

$$= \frac{67 \times 66 - 46 \times 52}{\sqrt{67 \times 102 - (46)^2}\sqrt{67 \times 116 - (52)^2}}$$

$$= \frac{4422 - 2392}{\sqrt{6834 - 2116}\sqrt{7772 - 2704}}$$

$$= \frac{2030}{\sqrt{4718}\sqrt{5068}}$$

$$= \frac{2030}{68.688 \times 71.19}$$

$$= \frac{2030}{4889.899} = 0.415.$$

Assumption of the Pearsonian Coefficient

Karl Pearson's coefficient of correlation is based on the following assumptions :

1. The two variables under study are affected by a larger number of independent causes so as to form a normal distribution. Variables like height, weight, price, demand, supply, etc., are affected by such forces that a normal distribution is formed.
2. There is linear relationship between the variables, *i.e.*, when the two variables are plotted on a scatter diagram a straight ling will be formed by the points so plotted.
3. There is a cause and effect relationship between the forces affecting the distribution of the items in the two series. If such a relationship is not formed between the variables, *i.e.*, if the variables are independent, there cannot be any correlation. For example, there is no relationship between income and height because the forces that affect these variables are not common.

Merits and Limitations of the Pearsonian Coefficient

However, the utility of this coefficient depends in part on a wide knowledge of the meaning of this 'yardstick' together with its limitations. The chief limitations of the method are :

1. As compared with other methods this method takes more time to compute the value of correlation coefficient.
2. The value of the coefficient is unduly affected by the extreme items.
3. Great care must be exercised in interpreting the value of this coefficient as very often the coefficient is misinterpreted.
4. The correlation coefficient always assumes linear relationship regardless of the fact whether that assumption is correct or not.

Interpreting Coefficient of Correlation

The full significance of r will only be grasped after working out a number of correlation problems and seeing the kinds of data that give rise to various values of r. The investigator must know his data thoroughly in order to avoid errors of interpretation and emphasis. He must be familiar, or become familiar, with all the relationships and theory which bear upon the data and should reach a conclusion based on logical reasoning and intelligent investigation on significantly related matters. However, the following general rules are given which would help in interpreting the value of r :

1. When r = – 1, it means there is perfect negative relationship between the variables.
2. When r = + 1, it means there is perfect positive relationship between the variables.
3. When r = 0, it means that there is no relationship between the variables, *i.e.*, the variables are uncorrelated.
4. The closer r is to + 1 or – 1, the closer the relationship between the variables and the closer r is to 0, the less close the relationship. Beyound this it is not safe to go. The full interpretation of r depends upon circumstances one of which is the size of the sample. All that can really be said that when estimating the value of one variable from the value of another, the higher the value of r the better the estimates.
5. The closeness of the relationship is not proportional to r. If the value of r is 0.8 it does not indicate a relationship twice as close as one of 0.4. It is, in fact, very much closer.

COEFFICIENT OF CORRELATION AND PROBABLE ERROR

The probable error of the coefficient of correlation helps in interpreting its value. With the help of probable error it is possible to determine the reliability of the value of the coefficient in so far as it depends on the conditions of random sampling. The probable error of the coefficient of correlation is obtained as follows :

$$P.E._r = 0.6745 \frac{1 - r^2}{\sqrt{N}}$$

where r is the coefficient of correlation and N the number of pairs of observations.

1. If the value of *r* is more than six times the probable error, the coefficient of correlation is practically certain, *i.e.* the value of *r* is significant.
2. If the value of *r* is less than the probable error there is no evidence of correlation, *i.e.*, the value of *r* is not at all significant.
3. By adding and substracting the value of probable error from the coefficient within which coefficient of correlation in the population can be expected to lie. Symbolically,

$$\rho = r \pm P.E.$$

where ρ (rho) denoted correlation in the population.

Let us compute probable error, assuming a coefficient of correlation of 0.80 and a sample of pairs of items. We will have

$$P.E. = 0.6745 \frac{1 - (.8)^2}{\sqrt{16}} = .06$$

The limits of the correlation in the population would be r ± P.E., *i.e.* 0.8 ± 0.06 or 0.74 – 0.86.

Instances are quite common wherein a correlation coefficient of 0.5 or even 0.4 is obviously considered to be a fairly high degree of correlation by a writer or research worker. Yet a correlation coefficient of 0.5 means that only 25 percent of the variation is explained. A correlation coefficient of 0.4 means that only 16 percent of the variation is explained.

Conditions for the Use of Probable Error

The measure of probable error can be properly used only when the following three conditions exist :

1. The statistical measure for which the P.E. is computed must have been calculated from a sample.
2. The data must approximate a normal frequency curve (bell-shaped curve).
3. The sample must have been selected in an unblased manner and the individual items must be independent.

However, these conditions are generally satisfied and as such the reliability of the correlation coefficient is determined largely on the basis of exterior tests of reasonableness which are often of a statistical character.

Example 4:

If r = 0.6 and N = 64, find out the probable error of the coefficient of correlation and determine the limits for population r.

Solution:

$$P.E._r = 0.6745 \frac{1 - r^2}{\sqrt{N}}$$

$$r = 0.6 \text{ and } N = 64$$

$$P.E._r = 0.64745 \frac{1 - (0.6)^2}{\sqrt{(64)}}$$

$$= \frac{0.6745 \times 0.64}{8} = 0.054.$$

Limits of population correlation = 0.6 ± 0.054
= 0.546 to 0.654.

COEFFICIENT OF DETERMINATION

One very convenient and useful way of interpreting the value of coefficient of correlation between two variables is to use square of coefficient of correlation, which is called coefficient of determination. The coefficient of determination thus equals r^2. If the value of r = 0.9, r^2 will be 0.81 and this would mean that 81 percent of the variation in the dependent variable has been explained by the independent variable. The maximum value of r^2 is unity because it is possible to explain all of the variation in Y, but it is not possible to explain more than all of it.

The coefficient of determination (r^2) is defined as the ratio of the explained variance to the total variance.

$$\text{Coefficient of determination} = \frac{\text{Explained Variation}}{\text{Total Variance}}$$

The ratio of explained variance to total variance is frequently called coefficient of non-determination. The coefficient of non-determination is denoted by K^2 and its square root is called the coefficient of alienation, or K. The K and K^2 value may also be used as the measure of the degree or relationship between two variables. For example, the higher the unexplained variance with respect to total variance, the higher will be the value of K^2 and the value of K. However, r^2 and r are more convenient in interpreting the result of correlation analysis.

It is much easier to understand the meaning of r^2 than *r* and, therefore, the coefficient of determination should be preferred in presenting the result of correlation analysis. Tuttle has beautifully pointed out that "the coefficient of correlation has been grossly overrated and is used entirely too much. Its square, the coefficient of determination, is a much more useful measure of the linear covariation of two variables. The reader should develop the habit of squaring every correlation coefficient he finds cited or stated before coming to any conclusion about the extent of the linear relationship between the two correlated variables."

The relationship between r and r^2 may be noted—as the value of r decreases from its maximum value of 1, the value of r^2 decreases much more rapidly, r will of course always be larger than r^2, unless $r^2 = 0$ or 1, 0 when $r = r^2$.

r	r^2	r	r^2
0.90	0.81	0.60	0.36
0.80	0.64	0.50	0.25
0.70	0.49	0.40	0.16

Thus the coefficient of correlation is 0.707 when just half the variance in Y is due to X.

It should be clearly noted that the fact that a correlation between two variables has a value of r = 0.60 and the correlation between two other variables has a value of r = 0.30 does not demonstrate that the first correlation is twice as strong as the second. The relationship between the two given values of r can better be understood by computing the value of r^2. When r = 0.06, r = 0.036 and when r = 0.30, r^2 = 0.90. This implies that in the first case 36% of the total variation is explained whereas in the second case 90% of the total variation is explained.

The coefficient of determination is a highly useful measure. However, it is often misinterpreted. The term itself may be misleading in that it implies that the variable X stands in a determining or causal relationship to the variable Y. The statistical evidence itself never establishes the existence of such causality. All that the statistical evidence can do is to define covariation, that term being used in a perfectly neutral sense. Whether causality is present or not and which way it runs if it is present, must be determined on the basis of evidence other than the quantitative observations.

However, r^2 is always a positive number. It cannot tell whether the relationship between the two variables is positive or negative. Thus, the square root of r^2 *i.e.*, $\sqrt{r^2} = \pm r$ is frequently computed to indicate the direction of the relationship, in addition to indicating the degree of relationship. Since the range of r^2 is from 0 to 1, the coefficient of correlation r will vary within the range of $\sqrt{0}$ to $\sqrt{1}$, or from ± 1. The + (plus) sign of *r* will indicate positive correlation, whereas the – (minus) sign will mean a negative correlation.

PROPERTIES OF THE COEFFICIENT OF CORRELATION

The following are the important properties of the correlation coefficient. *r*:

1. The coefficient of correlation lies between – 1 and + 1. Symbolically, $- \leq r \leq +1$ or $| r | \leq 1$.

Proof:

Let x and y be deviations of X and Y series from their mean and σ_x and σ_y be their standard deviations. Expand the function :

$$\Sigma = \left(\frac{x}{\sigma_x} + \frac{y}{\sigma_y}\right)^2 = \Sigma\left[\frac{x^2}{\sigma_x^2} + \frac{y^2}{\sigma_y^2} + \frac{2xy}{\sigma_x \sigma_y}\right]$$

$$= \frac{\Sigma x^2}{\sigma_x^2} + \frac{\Sigma y^2}{\sigma_y^2} + \frac{2\Sigma xy}{\sigma_x \sigma_y}$$

But $\dfrac{\Sigma x^2}{\sigma_x^2} = N \left[\because \sigma_x^3 = \dfrac{\Sigma x^2}{N} \text{ or } \dfrac{\Sigma x^2}{\sigma_x^2} = \dfrac{\Sigma x^2}{\Sigma x^2} \times N = N\right]$

Similarly, $\dfrac{y^2}{\sigma_y^2} = N$

Also, $= \dfrac{2xy}{\sigma_x \sigma_y} = 2Nr \qquad \left[\because r = \dfrac{\Sigma xy}{N\sigma_x \sigma_y}\right]$

Hence

$$\Sigma\left(\frac{x}{\sigma_x} + \frac{y}{\sigma_y}\right)^2 = N + N + 2Nr = 2N + 2Nr = 2N(1 + r).$$

But $\Sigma\left(\dfrac{x}{\sigma_x} + \dfrac{y}{\sigma_y}\right)$ is the sum of squares of real quantities and as such it cannot b negative; at the most it can be zero.

$\therefore \quad 2N (1 + r \geq 0)$

Hence r cannot be less than – 1; at the most it can be – 1.

Similarly, by expanding $\left(\frac{x}{\sigma_x} + \frac{y}{\sigma_y}\right)^2$ it can be shown that this is equal to 2 N (r – 1).

The again cannot be negative : at the most it can be zero.

$\because$ r cannot be grater than + 1 : at the most it can be + 1.

Hence $\quad -1 \leq r \leq +1$.

2. The coefficient of correlation is independent of change of scale and origin of the variable X and Y.

Proof:

By change of origin we mean subtracting some constant from every given value of X and Y and by change of scale we mean dividing or multiplying every value of X and Y by some constant.

We know that $r_{xy} = \dfrac{\Sigma(X - \overline{X})(Y - \overline{Y})}{\sqrt{\Sigma(X - \overline{X})^2 (Y - \overline{Y})^2}}$

where $\overline{X}$ and $\overline{Y}$ refer to actual means of X and Y series.

Let us now change the scale and origin. Deduct a fixed quantity a from X and b from Y. Also divide X and Y series by a fixed value i and c. After these changes are introduced, new values of x and y obtained from original X and Y shall be

$$x = \frac{X - a}{i} \text{ and } y = \frac{Y - b}{c}$$

$$\text{Mean of } x = \frac{\Sigma\left(\frac{(X - a)}{i}\right)}{n} = \frac{\frac{\Sigma X - Na}{i}}{N} = \frac{\Sigma X - Na}{Ni}$$

But $\quad \dfrac{\Sigma X - Na}{Ni} = \dfrac{X - a}{i}$

Thus, $\quad$ mean of $x = \dfrac{\overline{X} - a}{i}$

Similarly, it can be shown that mean of $y = \dfrac{\overline{Y} - b}{c}$

The value of coefficient of correlation r, for new set of values will be

$$r_{xy} = \frac{\Sigma\left(\frac{X-a}{i} - \frac{\overline{X}-a}{i}\right)\Sigma\left(\frac{Y-b}{c} - \frac{\overline{Y}-b}{c}\right)}{\sqrt{\Sigma\left(\frac{X-a}{i} - \frac{\overline{X}-a}{i}\right)^2 \Sigma\left(\frac{Y-b}{c} - \frac{\overline{Y}-b}{c}\right)^2}}$$

$$= \frac{\Sigma\left(\frac{X-a-\overline{X}-a}{i}\right)\Sigma\left(\frac{Y-b-\overline{Y}-b}{c}\right)}{\sqrt{\Sigma\left(\frac{X-a-\overline{X}-a}{i}\right)^2 \Sigma\left(\frac{Y-b-\overline{Y}-b}{c}\right)^2}}$$

$$= \frac{\dfrac{\Sigma(X-\overline{X})(Y-\overline{Y})}{ic}}{\sqrt{\dfrac{\Sigma(X-\overline{X})^2}{i^2} \times \dfrac{\Sigma(Y-\overline{Y})^2}{c^2}}}$$

$$= \frac{\Sigma(X-\overline{X})(Y-\overline{Y})/ic}{\sqrt{\Sigma(X-\overline{X})^2\,(Y-\overline{Y})^2/i^2c^2}} = \frac{\Sigma(X-\overline{X})(Y-\overline{Y})}{\sqrt{\Sigma(X-\overline{X})^2\,\Sigma(Y-\overline{Y})^2}}$$

Thus the coefficient of correlation is independent of change of scale and origin.

3. The coefficient of correlation is the geometric mean of two regression coefficient.

Symbolically, $r = \sqrt{b_{xy} \times b_{yx}}$

4. The degree of relationship between the two variables is symmetric as shown below :

$$r_{xy} = r_{yx}$$

$$r_{xy} = \frac{\Sigma xy}{N\sigma_x\sigma_y} = \frac{\Sigma yx}{N\sigma_y\sigma_x} = r_{yx}.$$

RANK CORRELATION COEFFICIENT

The Karl Pearson's method is based on the assumption that the population being studied is normally distributed. When it is known that the population is not normal or when the shape of the distribution is not known, there is need for a measure of correlation that involves no assumption about the parameter of the population.

It is possible to avoid making any assumptions about the populations being studied by ranking the observations according to size and basing the calculations on the ranks rather than upon the original observations. It does not matter which way the items are ranked, item number one may be the largest or it may be the smallest. Using ranks rather than actual observations gives the coefficient of rank correlation.

This method of finding out covariablity or the lack of it between two variables was developed by the British Psychologist Charles Edward Spearman in 1904. This measure is especially useful when quantitative measures for certain factors (such as in the evaluation of leadership ability or the judgment of female beauty) cannot be fixed, but the individual in the group can be arranged in order thereby obtaining for each individual a number indicating his (her) rank in the group. Spearman's rank correlation coefficient is defined as :

$$R = 1 - \frac{6\Sigma D^2}{N(N^2-1)} \text{ or } 1 - \frac{6\Sigma D^2}{N^3 - N}$$

where R denotes rank coefficient of correlation and D refers to the difference of rank between paired items in two series.

Consider a set of n individuals, ranked according to two characters X and Y.

Individual	$A_1\ A_2 \ldots A_1 \ldots A_n$
Ranking (X)	$x_1\ x_2 \ldots x_1 \ldots x_n$
Ranking (Y)	$y_1\ y_2 \ldots y_1 \ldots y_n$

We know that,

$$\bar{x} = \frac{1}{N}\Sigma x_1 = \frac{1}{X}[1 + 2 + 3 \ldots + N] = \frac{N+1}{2}$$

$$\bar{y} = \frac{1}{N}\Sigma y_1 = \frac{1}{Y}[1 + 2 + 3 \ldots + N] = \frac{N+1}{2}$$

So, $\bar{x} = \bar{y}$

$$\sigma_x^2 = \frac{N^2 - 1}{12} = \sigma_y^2 \text{ or } r = \frac{\mu_{11}}{\sigma_x \sigma_y}$$

Now if d_i stands for the difference in ranks of ith individual, we have

$$d_i = x_i = y_i = (x_i - x) - (y_i - y)$$

$$= x_i' - y_i'$$

where x_i' and y_i' are deviations of x_i and y_i for the mean $\bar{x}$ and $\bar{y}$

$$\Sigma d_i^2 = \Sigma (x_i' - y_i')^2$$

$$= \Sigma x_i'^2 + \Sigma y_i'^2 - 2\Sigma x_i' y_i'$$

$$\Sigma d_i^2 = \frac{1}{N}\Sigma x_i'^2 + \frac{1}{N}\Sigma y_i'^2 - \frac{2\Sigma s_i' y_i'}{N}$$
$$= \sigma_x^2 + \sigma_y^2 + 2\mu_{11}$$
$$= \sigma_x^2 + \sigma_y^2 - 2r\sigma_x\sigma_y$$
$$= 2\sigma_x^2 - 2r\sigma_x^2$$
$$= 2\sigma_x^2 (1-r)$$
$$\frac{1}{N} = \Sigma d_i^2 = 2\sigma_x^2 (1-r)$$
$$(1-r) = \frac{1}{N}\frac{\Sigma d_i^2}{2\sigma_x^2}$$
$$(1-r) = \frac{1}{N}\frac{\Sigma d_i^2}{\frac{2(N^2-1)}{12}} = \frac{1}{N}\frac{6\,\Sigma d_i^2}{(N^2-1)}$$
$$r = 1 - \frac{6\,\Sigma d_i^2}{N(N^2-1)} \text{ or } 1 - \frac{6\,\Sigma d_i^2}{N^3 - N}$$

The value of this coefficient interpreted in the same way as Karl Pearson's correlation coefficient, ranges between + 1 and – 1. When r_2 is + 1 there is complete agreement in the order of the ranks and the ranks are in the same direction. When r_2 is – 1 there is complete agreement in the order of the ranks and they are in opposite directions. This shall be clear from the following:

R_1	R_2	D $(R_1 - R_2)$	D^2	R_1	R_2	D $(R_1 - R_2)$	D^2
1	1	0	0	1	3	– 2	4
2	2	0	0	2	2	0	0
3	3	0	0	3	1	2	4
			$\Sigma D^2 = 0$				$\Sigma D^2 = 8$

$$R = 1 - \frac{6\,\Sigma D^2}{N^3 - N}$$
$$R = 1 - \frac{6\,\Sigma D^2}{N^3 - N}$$
$$= 1 - \frac{6 \times 0}{3^3 - 2} = 1 - 0 = 1$$

$$= 1 - \frac{6 \times 8}{3^3 - 2} = 1 - 2 = -1$$

Features of Spearman's Correlation Coefficient

1. The Spearman's correlation coefficient is nothing but Karl Pearson's correlation coefficient between the ranks. Hence, it can be interpreted in the same manner as Pearsonian correlation coefficient.
2. Spearman's correlation coefficient is distribution-free or non-parametric because no strict assumptions are made about the form of population from which sample observations are drawn.
3. The sum of the difference of ranks between two variables shall be zero. Symbolically. $\Sigma\ d = 0$.

In rank correlation we may have two types of problems :

(a) Where ranks are given.

(b) Where ranks are not given.

Where Ranks are Given

Where actual ranks are given to us the steps required for computing rank correlation are :

(i) Take the differences of the two ranks, i.e., $(R_1 - R_3)$ and denote these differences by D.

(ii) Square these differences and obtain the total $\Sigma\ D^2$.

(iii) Apply the formula $R = 1 - \frac{6\ \Sigma\ D^2}{N^3 - N}$.

Example 1:

Calculate the coefficient of correlation from the following data of the Spearman's. Rank difference method :

Price of Tea (Rs.)	*Price of Coffee (Rs.)*	*Price of Tea (Rs.)*	*Price of Coffee (Rs.)*
75	*120*	*60*	*110*
88	*134*	*80*	*140*
95	*15081*	*142*	
70	*115*	*50*	*100*

Solution:

Calculation of Spearman's Correlation Coefficient

Price of Tea (Rs.)	R_1	Price of Coffee (Rs.)	R_2	$(R_1 - R_2)^2$ D^2
75	4	120	4	0
88	7	134	5	4
95	8	150	8	0
70	3	115	3	0
60	2	110	2	0
80	5	140	6	1
81	6	142	7	1
50	1	100	1	0
				$\Sigma D^2 = 6$

$$R = 1 - \frac{6 \Sigma D^2}{N^3 - N} = 1 - \frac{6 \times 6}{8^3 - 8}$$

$$= 1 - \frac{36}{512 - 8}$$

$$= 1 - 0.071 = + 0.329 .$$

Example 2:

Quotations of Index Numbers of security prices of a certain joint stock company are given below :

Year	*Debenture price*	*Share price*
1	*97.8*	*73.2*
2	*99.2*	*85.8*
3	*98.8*	*78.9*
4	*98.3*	*75.8*
5	*98.4*	*77.2*
6	*96.7*	*87.2*
7	*97.1*	*83.8*

Using rank correlation method, determine the relationship between debenture prices and share prices.

Solution:

Calculation of Rank Correlation Coefficient

X	R_x	Y	R_y	$(R_x - R_y)^2$ D^2
97.8	3	93.2	1	4
99.2	7	85.8	6	1
98.8	6	78.9	4	4
98.3	4	75.8	2	4
98.4	5	77.2	3	4
96.7	1	87.2	7	36
97.1	2	83.8	5	9
				$\Sigma D^2 = 62$

$$R = 1 - \frac{6\,\Sigma D^2}{N^3 - N} = 1 - \frac{6 \times 62}{7^3 - 7}$$

$$= 1 - \frac{372}{336} = 1 - 1.107 = -0.107$$

Example 3:

Ten competitors is a beauty contest are ranked by three judges in the following order :

1st judge	*1*	*6*	*5*	*10*	*3*	*2*	*4*	*9*	*7*	*8*
2nd judge	*3*	*5*	*8*	*4*	*7*	*10*	*2*	*1*	*6*	*9*
3rd judge	*6*	*4*	*9*	*8*	*1*	*2*	*3*	*10*	*5*	*7*

Use the rank correlation coefficient to determine which pair of judges has the nearest approach to common tastes in beauty.

Solution:

In order to find out which pair of judges has the nearest approach to common tastes in beauty, we compare Rank Correlation between the judgments of :

(i) 1st judge and 2nd judge.

(ii) 2nd judge and 3rd judge, and

(iii) 1st judge and 3rd judge.

Rank by 1st judge R_1	Rank by 2nd judge R_2	Rank by 3rd judge R_3	$(R_1 - R_2)^2$ D^2	$(R_2 - R_3)^2$ D^2	$(R_1 - R_3)^2$ D^2
1	3	6	4	9	25
6	5	4	1	1	4
5	8	9	9	1	16
10	4	8	36	16	4
3	7	1	16	36	4
2	10	2	64	64	0
4	2	3	4	1	1
9	1	10	64	81	1
7	6	5	1	1	4
8	9	7	1	4	1
N = 10	N = 10	N = 10	$\Sigma D^2 = 200$	$\Sigma D^2 = 214$	$\Sigma D^2 = 60$

Rank correlation between the judgment of 1st and 2nd judges :

$$R = 1 - \frac{6\,\Sigma D^2}{N^3 - N}$$

Here we have directly calculated D^2 because D's are not required in applying the formula.

$$\therefore \quad R_{(\text{I and II})} = 1 - \frac{6 \times 200}{10^3 - 10}$$

$$= 1 - \frac{1200}{990} = 1 - 1.212 = -\,0.212$$

Rank correlation between the judgments of 2nd and 3rd judges :

$$R = 1 - \frac{6\,\Sigma D^2}{N^3 - N}$$

$$\therefore \quad R_{(\text{II and III})} = 1 - \frac{6 \times 214}{10^3 - 10} = 1 - 1.297 = -\,0.297$$

Rank correlation between judgments of 1st and 3rd judge :

$$R_{(\text{I and III})} = 1 - \frac{6\,\Sigma D^2}{N^3 - N}$$

$$= 1 - \frac{6 \times 60}{10^3 - 10} = 1 - \frac{360}{990} = 0.636.$$

Since coefficient is maximum in the judgments of the first and third judges we conclude that they have the nearest approach to common tastes in beauty.

Example 4:

Calculate Speaman's coefficient of correlation between marks assigned to ten students by judges X and Y in a certain competitive test as shown below:

S. No.	:	*1*	*2*	*3*	*4*	*5*	*6*	*7*	*8*	*9*	*10*
Marks by judge X	:	*52*	*53*	*42*	*60*	*45*	*41*	*37*	*38*	*25*	*27*
Marks by judge Y	:	*65*	*68*	*43*	*38*	*77*	*48*	*35*	*30*	*25*	*50*

Solution:

First assign ranks and then calculate rank correlation coefficient

Computation of Spearman's Coefficient of Correlation

Marks by judge X	R_x	**Marks by judge Y**	R_y	$(R_x - R_y)$ D^2
52	8	65	8	0
53	9	68	9	0
42	6	43	5	1
60	10	38	4	36
45	7	77	10	9
41	5	48	6	1
37	3	30	3	0
38	4	32	2	4
25	1	25	1	0
27	2	50	7	25
				$\Sigma D^2 = 76$

$$R = 1 - \frac{6\,\Sigma D^2}{N^3 - N} = 1 - \frac{6 \times 76}{10^3 - 10} = 1 - 0.461 = -\,0.539.$$

Example 5:

The ranking of 10 students in two subjects A and B are as follows :

A	*B*	*A*	*B*
6	*3*	*4*	*6*
5	*8*	*9*	*10*
3	*4*	*7*	*7*
10	*9*	*8*	*5*
2	*1*	*1*	*2*

Calculate rank correlation coefficient.

Solution:

Calculation of Rank Correlation Coefficient

R_1	R_2	$(R_1 - R_2)^2$ D^2
6	3	9
5	8	9
3	4	1
10	9	1
2	1	1
4	6	4
9	10	1
7	7	0
8	5	9
1	2	1
		$\Sigma D^2 = 36$

$$R = 1 - \frac{6\,\Sigma D^2}{N^3 - N} = 1 - \frac{6 \times 36}{10^3 - 10} = 1 - \frac{216}{990} = 0.782$$

Example 6:

Two ladies were asked to rank 7 different types of lipsticks. The ranks give by them are as follows :

Lipsticks	*A*	*B*	*C*	*D*	*E*	*F*	*G*
Neelu	*2*	*1*	*4*	*3*	*5*	*7*	*6*
Neera	*1*	*3*	*2*	*4*	*5*	*6*	*7*

Calculate Spearman's rank correlation coefficient.

Solution:

Calculation of Spearman's Rank Correlation Coefficient

X (R_1)	Y (R_2)	($R_1 - R_2$) D	D^2
2	1	+ 1	1
1	3	− 2	4
4	2	+ 2	4
3	4	− 1	1
5	5	0	0
7	6	+ 1	1
6	7	− 1	1
			$\Sigma D^2 = 12$

$$R = 1 - \frac{6\,\Sigma D^2}{N^3 - N} = 1 - \frac{6 \times 12}{7^3 - 7}$$

$$= 1 - 0.214 = 0.786.$$

Where Ranks are Not Given : When we are given the actual data and not the ranks. It will be necessary to assign the ranks. Ranks can be assigned by taking either highest value as I or the lowest value of 1. But whether we start with the lowest value or the highest value we must follow the same method in case of both the variables.

Equal Ranks

In some cases it may be found necessary to rank two or more individuals or entries as equal. In such a case it is customary to give each individual an average rank. Thus, if two individuals are ranked equal at fifth place, they are each given the rank $\frac{5+6}{2}$, that is 5.5 while. It three are ranked equal at fifth place, they are given the rank $\frac{5+6+7}{3} = 6$. In other words, where two or more items are to be ranked equal the rank assigned for purposes of calculating coefficient of correlation is the average of the ranks which these individuals would have gat had they differed slightly from each other.

Where equal ranks are assigned to some entries an adjustment in the above formula for calculating the rank coefficient of correlation is made.

The adjustment consists of adding $\frac{1}{12}(m^3 - m)$ to the value of ΣD^2, where M stands for the numbers of items whose ranks are common. If there are more than one such group of items with common rank, this value is added as many times the number of such groups. The formula can thus be written

$$\therefore R = 1 - \frac{6\left\{\Sigma D^2 + \frac{1}{12}(m^3 - m) + \frac{1}{12}(m^3 - m) + \ldots\right\}}{N^3 - N}$$

Example 7:

Obtain the rank correlation coefficient between the variables X and Y from the following pairs of observed values.

X :	*50*	*55*	*65*	*50*	*55*	*60*	*50*	*65*	*70*	*75*
Y :	*110*	*110*	*115*	*125*	*140*	*115*	*130*	*120*	*115*	*160*

Solution:

For finding ranks correlation coefficient first rank two various values. Taking lowest as 1 and next higher as 2 etc.

X	*Rank X* R_1	Y	*Rank Y* R_2	$(R_1 - R_2)$ D^2
50	2	110	1.5	0.25
55	4.5	110	1.5	9.00
65	7.5	115	4	12.25
50	2	125	7	25.00
55	4.5	140	9	20.25
60	6	115	4	4.00
50	2	130	8	36.00
65	7.5	120	6	2.25
70	9	115	4	25.00
75	10	160	10	00.00
				$\Sigma D^2 = 134$

It may be noted that in series X, 50 has repeated thrice (m = 3), 55 has been repeated twice (m = 2), 65 has been repeated twice (m = 2). In series Y, 110 has been repeated twice (m = 2) and 115 twice (m = 3).

$$R = 1 - \frac{6\left\{\Sigma D^2 + \frac{1}{12}(m^3 - m) + \frac{1}{12}(m^3 - m) + \frac{1}{12}(m^3 - m) + \frac{1}{12}(m^3 - m) + \frac{1}{12}(m^3 - m)\right\}}{N^3 - N}$$

$$R = 1 - \frac{6\left\{134 + \frac{1}{12}(3^3 - 3) + \frac{1}{12}(2^3 - 2) + \frac{1}{12}(2^3 - 2) + \frac{1}{12}(2^3 - 2) + \frac{1}{12}(3^3 - 3)\right\}}{10^3 - 10}$$

$$= 1 - \frac{6[134 + 2 + .5 + .5 + .5 + 2]}{990}$$

$$= 1 - 6\,\frac{[139.5]}{990} = 1 - \frac{837}{990} = 1 - 845 = 0.155$$

Merits and Limitations of the Rank Method

Merits

The merits of the Rank Method can be discussed here :

1. Even where actual data are given, rank method can be applied for ascertaining correlation.
2. This method is simpler to understand and eastern to apply compared to the Karl Pearson's method. The answers obtained by this method and the Karl Pearson's method will be the same provided no value is repeated, i.e., all the items are different.
3. This is the only method that can be used where we are given the ranks and not the actual data.
4. Where the data are of a qualitative nature like honesty, efficiency, intelligence, etc., this method can be used with great advantage. For example, the workers of two factories can be ranked in order of efficiency and the degree of correlation established by applying this method.

Limitations

The method is however associated with a few limitations also :

(a) Where the number of items exceeds 30 the calculations become quite tedious and require a lot of time. Therefore, this method should not be applied where N exceeds 30 unless we are given the ranks and not the actual values of the variables.

(b) This method cannot be used for finding out correlation in a grouped frequency distribution.

When to use Rank Correlation Coefficient:

The rank method has principal uses :

(a) The initial data are in the form of ranks.

(b) If N is fairly small (say, not more than 25 or 30) rank method is sometimes applied to interval data as an approximation to the more time consuming r. This requires that the interval data be transferred to rank orders for both variables. If N is much in excess of 30, the labour required in ranking the scores becomes greater the is justified by the anticipated saving of time through the rank formula.

CONCURRENT DEVIATION METHOD

This method of studying correlation is the simplest of all the methods. The only thing that is required under this method is to find out the direction of change of X variable and Y variable. The formula applicable is :

$$r_c = \pm\sqrt{\pm\left(\frac{2C - n}{n}\right)}$$

where r_c stands for coefficient of correlation by the concurrent method; C stands for the number of concurrent deviations or the number of positive signs obtained after multiplying D_x with D_y.

n = Number of pairs of observation compared.

Steps.

(a) Multiply D_x with D_y and determine the value of C, i.e., the number of positive sign.

(b) Find out the direction of change of X variable, i.e., as compared with the first value, whether the second value is increasing or decreasing or is constant. If it is increasing put a + sign; If it is decreasing put a – sign (minus) if it is constant put zero. Similarly, as compared to second value, find out whether the third value is increasing, decreasing or constant. Repeat the same process for other values. Denote this column by D_x.

(c) In the same manner as discussed above, find out the direction of change of Y variable and denote this column by D_y.

(d) Apply the above formula, i.e.,

$$r_c = \pm\sqrt{\pm\left(\frac{2C - n}{n}\right)}$$

Note. The significance of ± signs, both (inside the under-root and outside the under-root, is that we cannot take the under-root of minus sign. Therefore, if $\frac{2C - n}{n}$ is negative, this negative value multiplied with the minus sign inside would make it positive and we can take the under-root. But the ultimate result would be negative. If $\frac{2C - n}{n}$ is positive then, of course, we get a positive value of the coefficient of correlation.

Example 1:

The ranks of the same 15 students in two subjects A and B are given below. The two numbers within brackes denote the ranks of the same student in A and B respectively.

(1, 10), (2, 7), (3, 2), (4, 6), (5, 4), (6, 8), (7, 3), (8, 1), (9, 1), (10, 15), (11, 19), (12, 5), (13, 14), (14, 12), (15, 13).

Find the Spearman's Rank Correlation Coefficient.

Solution:

Calculation of Spearman's Rank Correlation Coefficient

R_A	R_B	$(R_A - R_B)^2$ D^2
1	10	81
2	7	25
3	2	1
4	6	4
5	4	1
6	8	4
7	3	16
8	1	81
9	11	4
10	15	25
11	9	4
12	5	49
13	14	1
14	12	4
15	13	4
		$\Sigma D^2 = 304$

$$R = 1 - \frac{6 \Sigma D^2}{N^3 - N}$$

$$= 1 - \frac{6 \times 304}{15^3 - 15}$$

$$= 1 - \frac{1824}{3360} = 1 - 0.543 = 0.457.$$

Example 2:

The following data relate to age of employees and the number of days they were reported sick in a month.

	Employees									
	1	*2*	*3*	*4*	*5*	*6*	*7*	*8*	*9*	*10*
Age (X)	*30*	*32*	*35*	*40*	*48*	*50*	*52*	*55*	*57*	*61*

Sick days (Y) 1 0 2 5 2 4 6 5 7 8

Calculate Karl Pearson's Coefficient of Correlation and interpret it.

Solution:

Calculation of Karl Pearson's Coefficient of Correlation

Age X	$(X - \overline{X})$ x	x^2	Sick days Y	$(Y - \overline{Y})$ y	y^2	xy
30	– 16	256	1	– 3	9	+ 48
32	– 14	196	0	– 4	16	+ 56
35	– 11	121	2	– 2	4	+ 22
40	– 6	36	5	+ 1	1	– 6
48	+ 2	4	2	– 2	4	– 4
50	+ 4	16	4	0	0	0
52	+ 6	36	6	+ 2	4	+ 12
55	+ 9	81	5	+ 1	1	+ 9
57	+ 11	121	7	+ 3	9	+ 33
61	+ 15	225	8	+ 4	16	+ 60
$\Sigma X = 460$	$\Sigma x = 0$	$\Sigma x^2 = 1092$	$\Sigma Y = 40$	$\Sigma y = 0$	$\Sigma y^2 = 64$	$\Sigma xy = 230$

$$\overline{X} = \frac{460}{10} = 46, \ \overline{Y} = \frac{40}{10} = 4$$

Since the actual means of X and Y are not in fraction, we can use the direct method of calculating correlation coefficient.

$$r = \frac{\Sigma\, xy}{\sqrt{\Sigma\, x^2 \times \Sigma\, y^2}} = \frac{230}{\sqrt{1092 \times 64}}$$

$$= \frac{230}{263.363} = +0.87.$$

There is a high degree of positive correlation between age and number of days reported sick.

Example 3:

Find the coefficient of the correlation for the following :

Cost :	*39*	*65*	*62*	*90*	*82*	*75*	*25*	*98*	*36*	*78*
Sales :	*47*	*53*	*58*	*86*	*62*	*68*	*60*	*91*	*51*	*84*

Solution:

Calculation of Coefficient of Correlation By Karl Pearson's Method

X	$(X - \bar{X})$ $\bar{X} = 65$ x	x^2	Y	$(Y - \bar{Y})$ $\bar{Y} = 66$ y	y^2	xy
39	– 26	676	47	– 19	361	+ 494
65	0	0	53	– 13	169	0
62	– 3	9	58	– 8	64	+ 24
90	+ 25	625	86	+ 20	400	+ 500
82	+ 17	289	62	– 4	16	– 68
75	+ 10	100	68	+ 2	4	+ 20
25	– 40	1600	60	– 6	36	+ 240
98	+ 33	1089	91	+ 25	625	+ 825
36	– 29	841	51	– 15	225	+ 435
78	+ 13	169	84	+ 18	324	+ 234
$\Sigma X = 650$	$\Sigma x = 0$	$\Sigma x^2 = 5398$	$\Sigma Y = 660$	$\Sigma y = 0$	$\Sigma y^2 = 2224$	$\Sigma xy = 2704$

$$r = \frac{\Sigma xy}{\sqrt{\Sigma x^2 \times \Sigma y^2}}$$

$$= \frac{2704}{\sqrt{5398 \times 2224}} = \frac{2704}{3464.85} = +0.78.$$

Example 4:

With the following data in 6 cities calculate the coefficient of correlation by Pearson's method between the density of population and death rate :

City	*Area in kilometres*	*Population in 1000*	*No. of deaths*
A	*150*	*30*	*300*
B	*180*	*90*	*1440*
C	*100*	*40*	*560*
D	*60*	*42*	*840*
E	*120*	*72*	*1224*
F	*80*	*24*	*312*

Solution:

First we will calculate density of population and death rate and denote them by X and Y.

$$\text{Density} = \frac{\text{Population}}{\text{Area}}; \ \text{Death rate} = \frac{\text{No. of Deaths}}{\text{Population}} \times 1000$$

City	Density X	(X – 450)/100 x	x^2	Death Y	(Y – 15) y	y^2	xy
A	200	– 2.5	6.25	10	– 5	25	+ 12.5
B	500	+ 0.5	0.25	16	+ 1	1	+0.5
C	400	–0.5	0.25	14	– 1	1	+0.5
D	700	+ 2.5	6.25	20	+ 5	25	+ 12.5
E	600	+ 1.5	2.25	17	+ 2	4	+ 3.0
F	300	– 1.5	2.25	13	– 2	4	+ 3.0
		$\Sigma x = 0$	$\Sigma x^2 = 17.5$	$\Sigma Y = 90$	$\Sigma y = 0$	$\Sigma y^2 = 60$	$\Sigma xy = 32$

$$r = \frac{\Sigma xy}{\sqrt{\Sigma x^2 \times \Sigma y^2}}$$

$$= \frac{32}{\sqrt{17.5 \times 60}}$$

$$= \frac{32}{32.404} = +0.988.$$

Example 5:

A correlation coefficient of 0.5 does not mean that 50% of the data are explained. Comment.

Solution:

The coefficient of correlation does not indicate the percentage of explained variation. For correctly interpreting the correlation coefficient we have to find the value of coefficient of determination. The coefficient of determination is found by squaring r and it gives an idea determination. The coefficient of determination is found by squaring r and it gives an idea as to what proportion of variation of Y is explained by the variation in X. Thus $(0.5)^2$ or this does not mean that 25% of the data are explained, it only means that out of total variation of Y only 25.5 is due to X and the rest is due to other factors.

Example 6:

Calculate Pearson's coefficient of correlation from the following data using 44 and 26 as the origin of X and Y respectively :

X :	43	44	46	40	44	42	45	42	38	40	42	57
Y :	29	31	19	18	19	27	27	29	41	30	26	10

Solution:

Calculation of Pearson's Coefficient of Correlation

X	(X – 44) d_x	d_x^2	Y	(Y – 26) d_y	d_y^2	$d_x d_y$
43	– 1	1	29	+ 3	9	– 3
44	0	0	31	+ 5	25	0
46	+ 2	4	19	– 7	49	– 18
40	– 4	16	18	– 8	64	+ 32
44	0	0	19	– 7	49	– 14
42	– 2	4	27	+ 1	1	2
45	+ 1	1	27	+ 1	1	+ 1
42	– 2	4	29	+ 3	9	– 6
38	– 6	36	41	+ 15	225	– 90
40	– 4	16	30	+ 4	16	– 16
42	– 2	4	26	0	0	0
57	+ 13	169	10	– 16	256	– 208
	$\Sigma d_x = -5$	$\Sigma d_x^2 = 225$		$\Sigma d_y = -6$	$\Sigma d_y^2 = 704$	$\Sigma d_x d_y = 306$

$$r = \frac{N\Sigma d_x d_y - \Sigma d_x d_y}{\sqrt{N\Sigma d_x^2 - (\Sigma dx)^2}\sqrt{N\Sigma d_y^2 - (\Sigma d_y)^2}}$$

$$= \frac{12(-306) - (-5)(-6)}{\sqrt{12 \times 225 - (-5)^2}\sqrt{12 \times 704 - (-6)^2}}$$

$$= \frac{-3672 - 30}{\sqrt{3080 - 25}\sqrt{8448 - 36}} = \frac{-3702}{\sqrt{3035}\sqrt{8412}}$$

$$= \frac{-3702}{55.09 \times 91.72} = \frac{-3702}{5052.85} = -0.733.$$

Example 7:

Calculate the coefficient of correlation using the method of concurrent deviation between supply and demand of an item for a period of 10 years as given below :

Year	*1990*	*1991*	*1992*	*1993*	*1994*	*1995*	*1996*	*1997*	*1998*	*1999*
Supply	*125*	*160*	*164*	*174*	*155*	*170*	*165*	*162*	*172*	*175*
Demand	*225*	*125*	*192*	*190*	*165*	*174*	*124*	*127*	*152*	*169*

Solution:

Calculation of Correlation by Concurrent Deviation Method

Year	**X**	**Direction of change of X D_x**	**Y**	**Direction of change of Y D_y**	**D_xD_y**
1990	125		115		
1991	160	+	125	+	+
1992	164	+	192	+	+
1993	174	+	190	−	−
1994	155	−	165	−	+
1995	170	+	174	+	+
1996	165	−	124	−	+
1997	162	−	127	+	−
1998	172	+	152	+	+
1999	175	+	169	+	+
					C = 7

$$r_c = \pm\sqrt{\pm\frac{2l - n}{n}} = \pm\sqrt{\pm\frac{2 \times 7 - 9}{9}} = \sqrt{\frac{5}{9}} = 0.745$$

Merits and Limitations of Concurrent Deviation Method

Merits

The following are the basic advantages of this method:

(a) When the number of items is very large, this method may be used to form a quick idea about the degree of relationship before making use of more complicated methods.

(b) It is simplest of all the methods.

Limitations

This method is associated with the following limitations :

(a) The results obtained by this method are only a rough indicator of the presence or absence of correlation.

(b) This method does not differentiate between small and big changes. For example, if X increases from 100 to 101 the sign will be plus and if Y increases from 60 to 160, the sign will be plus. Thus, both get equal weight when they vary in the same direction.

Calculation of Correlation in Time Series

When we observe numerical data in relation to time the set of observations so obtained is known as time series. Time series depict two types of fluctuations: (i) long term and (ii) short term. While studying correlation between two time series, it is necessary to study separately the correlation of long-term changes and short-term changes. The reason is that the relationship between the long-term changes of two series may be quite different from that between the short-term change of these series. It is quite likely that in two time series there may be negative correlation between vice versa. Hence, it becomes necessary to study separately correlation between long-term changes and short-term changes as otherwise misleading results may be obtained.

Correlation of Long-term Changes

For finding out the correlation of long-term changes, the only thing required is to determine trend values for both the series by the moving average method or the method of least squares. After determining these trend values correlation can be obtained by the methods discussed above and no special method is necessary. The only thing to remember is that the coefficient of correlation shall be computed of the trend value of the two series and not of the original data.

Example 8:

Calculate coefficient of concurrent deviation from the following data :

Price	*Imports*	*Price*	*Imports*
368	*22*	*384*	*26*
384	*21*	*395*	*24*
385	*24*	*403*	*29*
361	*20*	*400*	*28*
347	*22*	*385*	*27*

Solution:

Calculation of Coefficient of Concurrent Deviation

Price X	Direction of change of variable X D_x	Imports Y	Direction of change of variable Y D_y	$D_x D_y$
368		22		
384	+	21	−	−
385	+	24	+	+
361	−	20	−	+
347	−	22	+	−
384	+	26	+	+
395	+	24	−	−
403	+	29	+	+
400	−	28	−	+
384	−	27	−	+
				C = 6

$$r_c = \pm\sqrt{\pm\left(\frac{2\,C - n}{n}\right)};\ C = 6,\ n = 9$$

$$r_c = \pm\sqrt{\pm\left(\frac{2 \times 6 - n}{9}\right)} = \pm\sqrt{\pm .333} = +0.577.$$

Example 9:

Calculate the Karl Pearson's coefficient of correlation of the short-term oscillations for the indices of supply and price of certain commodity given here :

Year	*Index of Supply*	*Index of Price*	*Year*	*Index of Supply*	*Index of Price*
1984	*91*	*117*	*1992*	*104*	*77*
1985	*98*	*97*	*1993*	*98*	*93*
1986	*95*	*102*	*1994*	*100*	*89*
1987	*92*	*108*	*1995*	*108*	*83*
1988	*93*	*106*	*1996*	*116*	*78*
1989	*95*	*96*	*1997*	*114*	*84*
1990	*102*	*77*	*1998*	*111*	*95*
1991	*107*	*68*			

(Take 5-year moving average and ignore decimals in computing the average)

$$r = \frac{\Sigma xy}{\sqrt{\Sigma x^2 \times \Sigma y^2}}$$

Solution:

Calculation of Coefficient of Correlation of Short-term Oscillations for the Indices of Supply and Price

Year	Index of Supply	5-yearly moving average X	Daviation of actual values from moving X	Square of deviations x^2	Index of price	5-yearly moving average Y	Deviation of actual values from moving Y	Square of deviations Y^2	Product of deviations xy
1984	91	—	—	—	117		—	—	—
1985	98	—	—	—	97	—	—	—	—
1986	95	93.8	+ 1.2	1.44	102	105.8	- 3.8	14.44	- 4.56
1987	92	94.8	- 2.8	7.84	108	101.6	+ 6.4	40.96	- 17.92
1988	93	98.0	- 2.6	6.76	105	97.6	+ 7.4	54.76	- 19.24
1989	96	98.0	- 2.0	4.00	96	90.8	+ 5.2	27.04	- 10.40
1990	102	100.4	+ 1.6	2.56	77	84.6	- 7.6	57.76	- 12.16
1991	107	101.3	+ 5.6	31.36	68	82.2	- 14.2	201.64	- 79.52
1992	104	102.2	+ 1.8	3.24	77	80.2	- 3.8	14.44	- 6.84
1993	98	103.4	- 5.4	20.16	93	82.0	+ 11.0	121.00	- 59.40
1994	100	105.2	- 5.2	27.04	89	84.0	+ 5.0	25.00	- 26.00
1995	108	107.2	+ 0.8	0.64	8393	85.4	- 2.4	5.76	- 1.92
1996	116	109.8	+ 6.2	38.44	78	85.4	- 7.4	54.76	- 45.88
1997	114	—	—	—	84	—	—	—	—
1998	111	—	—	—	95	—	—	—	—
				$\Sigma x^2 =$ 152.48				$\Sigma y^2 =$ 617.56	$\Sigma xy =$ - 283.84

$\Sigma xy = 283.84$, $\Sigma x^2 = 152.48$, $\Sigma y^2 = 617.56$

Substituting the values, $r = \dfrac{-283.56}{\sqrt{152.48 \times 617.56}}$

$$\log r = \log 283.84 - \frac{1}{2} [\log 152.48 + \log 617.56]$$

$$= 2.4531$$

$$- \frac{1}{2} [2.1831 + 2.7907] = 2.4531$$

$$= \frac{1}{2}[4.9738] = 2.4531$$

$$= 2.4869 = 1.9662.$$

$$r = \text{A L } 1.9662 = 0.925;$$

Hence $r = -0.925.$

Example 10:

Calculate the coefficient of concurrent deviation from the following :

X :	*60*	*55*	*50*	*56*	*30*	*70*	*40*	*35*	*80*	*80*	*75*
Y :	*65*	*40*	*35*	*75*	*63*	*80*	*35*	*20*	*80*	*60*	*50*

Solution:

Calculation of Correlation by Concurrent Deviation Method

X	D_x	Y	D_y	$D_x D_y$
60	–	65	–	+
55	–	40	–	+
50	–	35	–	+
56	+	75	+	+
30	–	63	–	+
70	+	80	+	+
40	–	35	–	+
35	–	20	–	+
80	+	80	+	+
80	0	60	–	0
75	–	50	–	+
				C = 8

$$r_c = \pm\sqrt{\pm\frac{2C - n}{n}}$$

$$= \pm\sqrt{\pm\frac{2 \times 8 - 10}{10}} = \sqrt{\frac{6}{10}} = 0.774$$

Correlation can be calculated by following steps.

(a) Deduct from the actual values the corresponding trend values obtained in step (i). This would give the short-term fluctuations. Devote these short-term fluctuations by the symbol x for X series and y for Y series.

(b) Square the short-term fluctuations of X series and obtain the total Σx^2.

(c) Square the short-term fluctuations of Y series and obtain the total Σy^2.

(d) Multiply x with y for each value and obtain the total Σ x y.

(e) Determine the trend value s by the moving average method.

(f) None apply the formula

$$r = \frac{\Sigma xy}{\sqrt{\Sigma x^2 \times \Sigma y^2}}$$

Here, x denotes deviation of X series from moving average and not from arithmetic mean. Similarly, y denotes deviation of Y series from moving average and not from arithmetic mean.

Thus, we find that the only difference between correlation explained earlier and correlation in short-term changes is that whereas in the former we take deviations from arithmetic mean, in the latter we take deviations from the trend values.

Special care must be taken while interpreting a significant correlation in time series as it is very likely that the correlation is the result of other variables which influence the correlated variable. A high degree of correlation may be observed between any two economic time series and this may be due to the remarkable and fairly steady growth of the economy. Yule states that the correlation between radio-receiving licences granted in the United Kingdom and number of notified mental defectives per 10,000 population from 1924 to 1937, inclusive, is very high (r = 0.998). This certainly seems to the person who attempts to explain it rationally as cause and effect, to be indeed a "nonsense correlation", as Yule calls such cases. Karl Pearson called such cases "Spurious Correlation". Yule has nicely explained that the two series were increasing through time, probably from different underlying causes although the growth of technology including medicine and mental health makes the very valid point that any two time series which are either rising of falling steadily (one may be rising, the other falling) will show negative or positive correlation, no matter what may be the reasons for the variation in the two series. This clearly shows that correlation means little except that the two series are varying in unison. If the causes are different fro the two series, the correlation may not continue in future.

In some cases the two correlated time series may not only be acted upon by other variables, but also one or both of them may influence the other. For example, one may find a high correlation between industrial profits and the wages of the workers most of the time. A logical argument can be made that each has some influence on the other but both are greatly influenced by the underlying variables which cause growth and prosperity (or recession in the economy).

LEG AND LEAD IN CORRELATION

The study of lag and lead is of special significance while studying economic and business series. In the correlation of time series the investigator may find that there is a time gap before a cause and effect relationship is established. For example, the supply of a commodity may increase today, but it may not have an immediate effect on prices—it may take a few days or even months for prices to adjust to the increased supply. This difference in the period before a cause and effect relationship is established is called 'Lag'. While computing correlation their time gap must be considered : otherwise fallacious conclusions may be drawn. The partring of items is adjusted according to the time lag.

Month	Supply	Price
Jan.	100	70
Feb.	105	69
March	108	80
April	12	72
May	118	75
June	120	70
July	125	74
Aug.	104	76
Sep.	122	78
Oct.	116	80
Nov.	122	78
Dec.	127	75

Taking the new pairs of values, correlation can be calculated in the same manner as discussed earlier.

Example 1:

Calculate Karl Pearson's Coefficient of Correlation between age and playing habits from the data given below. Also calculate probable error and comment on the value :

Age	:	*20*	*21*	*22*	*23*	*24*	*25*
No. of Students	:	*500*	*400*	*300*	*240*	*200*	*160*
Regular Players	:	*400*	*300*	*180*	*96*	*60*	*24*

Solution:

Let us first find the percentage of regular players and then calculate correlation between age and percentage.

Age	(X – 22) d_x	d_x^2	No. of Students	Regular Players	% or Regular Players (Y)	(Y – 50) d_y	d_y^2	$d_x d_y$
20	– 2	4	500	400	80	+ 30	900	– 60
21	– 1	1	400	300	75	+ 25	625	– 25
22	0	0	300	180	60	+ 10	100	0
23	+ 1	1	240	96	40	– 10	100	– 10
24	+ 2	4	200	60	30	– 20	400	– 40
25	+ 3	9	160	24	15	– 35	1225	– 105
Σ X = 135	Σ d_x = 3	Σ d_x^2 = 19			Σ Y = 300	Σ d_y = 0	Σ d_y^2 = 3350	Σ $d_x d_y$ = – 240

$$r = \frac{N\Sigma d_x d_y - \Sigma d_x d_y}{\sqrt{N\Sigma d_x^2 - (\Sigma dx)^2}\sqrt{N\Sigma d_y^2 - (\Sigma d_y)^2}}$$

$$= \frac{(6 \times -240) - (3 \times 0)}{\sqrt{6 \times 19 - (3)^2}\sqrt{6 \times 3350}}$$

$$= \frac{-1440}{\sqrt{105 \times 20100}} = \frac{-1440}{1452.756} = -0.991$$

$$P.E.r = 0.6745\frac{1 - r^2}{\sqrt{N}}$$

$$= 0.6745\frac{1 - (.991)^2}{\sqrt{6}}$$

$$= \frac{0.6745 \times .018}{2.449}$$

$$= 0.005.$$

Example 2:

The following are the monthly figures of advertising expenditure and sales of a firm. It is generally found that advertising expenditure has its impact on sales generally after 2 months. Allowing for this time lag, calculate coefficient of correlation.

Months	*Advertising Expenditure*	*Sales*
Jan	*50*	*1,200*
Feb.	*60*	*1,500*
March	*70*	*1,600*
April	*90*	*2,000*
May	*120*	*2,200*
June	*150*	*2,500*
July	*140*	*2,400*
Aug.	*160*	*2,600*
Sep.	*170*	*2,800*
Oct.	*190*	*2,900*
Nov.	*200*	*3,100*
Dec.	*250*	*3,900*

Solution:

Allow for a time lag of 2 months, i.e., link advertising expenditure of January with sales for March, and so on.

Calculation of Correlation Coefficient

Months	**Advertising Expenditure X**	**$\frac{(X-\overline{X})}{100}$ x**	**x^2**	**Sales Y**	**$\frac{(Y-\overline{Y})}{100}$ y**	**y^2**	**xy**
Jan.	50	– 7	49	1,600	– 10	100	70
Feb.	60	– 6	36	2,000	– 6	36	36
March	70	– 5	25	2,200	– 4	16	20
April	90	– 3	9	2,500	– 1	1	3
May	120	0	0	2,400	– 2	4	0
June	150	+ 3	9	2,600	0	0	0
July	140	+ 2	4	2,800	+ 2	4	4
Aug.	160	+ 4	16	2,900	+ 3	9	12
Sep.	170	+ 5	25	3,100	+ 5	25	25
Oct.	190	+ 7	49	3,900	+ 13	169	91
	$\Sigma X =$ 1,200	$\Sigma x = 0$	$\Sigma x^2 =$ 222	$\Sigma Y =$ 26,000	$\Sigma y = 0$	$\Sigma y^2 =$ 364	$\Sigma xy =$ 261

$$r = \frac{\Sigma xy}{\sqrt{\Sigma x^2 \times \Sigma y^2}}$$

$$\bar{x} = \frac{1{,}200}{100} = 120,$$

$$\bar{y} = \frac{26{,}000}{10} = 2{,}600$$

$$\Sigma xy = 261,\ \Sigma x^2 = 222,\ \Sigma y^2 = 364$$

$$r = \frac{261}{\sqrt{222 \times 364}} = \frac{261}{284.267}$$

$$= +\ 0.918.$$

Example 3:

If covariance between X and Y variables is 10 and the variances of X and Y are respectively 16 and 9, find the coefficient of correlation.

Solution:

$$\text{Covariance of X and Y} = \frac{\Sigma xy}{N} = 10$$

Variance of X = 16 $\quad \therefore \sigma_x = \sqrt{16} = 4$

Variance of Y = 9 $\quad \therefore \sigma_y = \sqrt{9} = 3$

$$r = \frac{\Sigma xy}{N \sigma_x \sigma_y}$$

Substituting the value; $r = \dfrac{10}{4 \times 3} = +0.833.$

Example 4:

Find the coefficient of correlation with the help of Karl Pearson's Method

Marks in Mathematics

		10	*20*	*30*	*40*	*50*
Marks in	*5*	*2*	*4*	*1*	*4*	*1*
Statistics	*10*	*8*	*2*	*5*	*1*	—
	15	—	*3*	*2*	*1*	—
	20	—	*1*	*3*	*2*	*4*
	25	—	—	*4*	*2*	—

Solution:

Let marks in Mathematics be denoted by X and the marks in Statistics by Y.

Computation of Correlation Coefficient

Y	X	10	20	30	40	50				
d_x / Y / d_y	-2	-1	0	$+1$	$+2$	f	fd_y	fd_y^2	fd_xd_y	
5	-2	+8 2	+8 4	0 1	-8 4	-4 1	12	-24	48	+4
10	-1	+16 8	+2 2	0 5	-1 1	—	18	-16	16	+17
15	0	—	3	0 2	0 1	—	6	0	0	0
20	+1	—	-1 1	0 3	+2 2	+8 4	10	+10	10	+9
25	+2	—	—	0 4	4 2	—	6	+12	24	+4
Total	2	10	10	15	10	5	N = 50	Σfd_y = -18	Σfd_y^2 = -98	Σfd_xd_y = 34
	fd_x	-20	-10	0	+10	+10	Σfd_x = -10			
	fd_x^2	40	10	0	10	20	Σfd_x^2 = 80			
	fd_xd_y	+24	+9	0	-3	+4	Σfd_xd_y = 34			

$$r = \frac{N\Sigma f d_x d_y - \Sigma f d_x \Sigma f d_y}{\sqrt{N\Sigma f d_x^2 - (\Sigma f dx)^2}\,\sqrt{N\Sigma f d_y^2 - (\Sigma f d_y)^2}}$$

$$= \frac{50 \times 34 - (-10)(-18)}{\sqrt{50 \times 80\,(-10)^2}\,\sqrt{98 \times 50 - (-18)^2}}$$

$$= \frac{1700 - 180}{\sqrt{3900}\,\sqrt{4900 - 324}}$$

$$= \frac{1520}{4224.499} = 0.36.$$

Example 5:

The coefficient of correlation between two variates X and Y 0.64. Their covariance is 16. The variance of X is 19. Find the standard deviation of Y series.

Solution:

Covariance $= \frac{\Sigma\, xy}{N} \quad \sigma = \sqrt{9} = 3$

$$r = \frac{\Sigma xy}{N\sigma_x \sigma_y} = \frac{\Sigma\, xy}{N} \times \frac{1}{\sigma_x \sigma_y} \quad \frac{\Sigma\, xy}{N} = 16,\ \sigma_x = 3,\ r = 64.$$

Substituting the values $64 = \frac{16}{3\sigma_y}$

$$1.92\ \sigma_y = 16$$

or $$\sigma_y = \frac{16}{1.92} = 8.333.$$

Hence the standard deviation of Y series is 8.333.

Example 6:

Coefficient of correlation between two variates X and Y is 0.3. The covesriance is 9. The variance of X is 16. Find the standard deviation of Y series.

(b) The coefficient of the rank correlation between debenture prices and share prices is found to be 0.143. If the sum of the squares of the differences in ranks is given to be 48, find the value of N.

Solution:

(a) Covariance $= \frac{\Sigma\, xy}{N}$, where x and y are the deviation of the values of X and Y series from their respective means.

Variance of X series is 16, i.e.

$$\sigma = \sqrt{16} = 4 \quad r = \frac{\Sigma xy}{N\sigma_x \sigma_y} = \frac{\Sigma\, xy}{N} \times \frac{1}{\sigma_x \times \sigma_y}$$

Substituting the given values, we get

$$0.3 = 9 \times \frac{3}{4\sigma_y}$$

$$1.2\ \sigma_y = 9 \text{ or } \sigma_y = \frac{9}{1.2} = 7.5$$

(b) $R = 1 - \frac{6\Sigma D^2}{N^3 - N}$; Here R = 0.143, $\Sigma\ D^2 = 48$

$$0.143 = 1 - \frac{6 \times 48}{N^3 - N} \text{ or } \frac{288}{N^3 - N} = 0.857$$

$$0.857\ (N^3 - N) = 288 \text{ or } (N^3 - N) = \frac{288}{0.857} = 336$$

or $\quad N^3 - N - 336 = 0$ or $N^3 - N - 343 + 7 = 0$

$$(N - 7)\ N^2 + 7\ N\ (N - 7) + 48\ (N - 7) = 0$$

or $$(N - 7)\ (N^2 - 7N + 48) = 0$$

either $$N - 7 = 0, \text{ i.e., } N = 7$$

or $$N^2 + 7N + 48 = 0$$

Since $b^2 - 4\ ac$ is negative, values of N belong to the set of complex numbers.

Example 7:

A student calculates the value of r as + 0.72 of a question comparison g5 pairs of observations and concludes that there is high degree of correlation between the variables. Do you justify hiss conclusion?

Solution:

The value of r can be interpreted by calculating probable error. If it is more the 6 times the probable error, it is considered significant :

$$P.E._r = .6745\ \frac{1 - r^2}{\sqrt{N}} = .6745\ \frac{1 - (.72)^2}{\sqrt{N}} = \frac{.6745 \times .4816}{.236} = 0.145$$

$$\frac{r}{P.E._r} = \frac{.72}{.145} = 4.97$$

Since r is less than 6 times the probable error it cannot be regarded as significant.

Example 8:

A computer while calculating the correlation coefficient between two variables X and Y from 25 pairs of observations obtained the following results :

$$n = 25,\ \Sigma X = 125,\ \Sigma X^2 = 650$$

$$\Sigma Y = 100,\ \Sigma Y^2 = 460,\ \Sigma XY = 508.$$

It was however, discovered at the time of checking that he had copied down two pairs as

X	*Y*
6	14
8	6

While the correct value was

X	*Y*
8	*12*
6	*8*

Obtain the correct value of correlation coefficient.

Solution:

Correct $\Sigma X = 125 - 6 - 8 + 8 + 6 = 125$

Correct $\Sigma X = 100 - 14 - 6 + 12 + 8 = 100$

Correct $\Sigma X^2 = 650 - (6)^2 - (8)^2 + (8)^2 + (6)^2$

$= 650 - 36 - 64 + 64 + 36 = 650$

Correct $\Sigma X^2 = 460 - (14)^2 - (6)^2 + (12)^2 + (8)^2$

$= 460 - 196 - 36 + 144 + 64 = 436$

Correct $\Sigma X Y = 508 - (6 \times 14) - (8 \times 6) + (8 \times 12) + (6 \times 8)$

$= 508 - 84 - 48 + 96 + 48 = 520$

$$r = \frac{\Sigma XY - N\overline{X}\,\overline{Y}}{\sqrt{\Sigma X^2 - N(\overline{X})^2}\,\sqrt{\Sigma Y^2 - N(\overline{Y})^2}}$$

$$\overline{X} = \frac{\Sigma X}{N} = \frac{125}{25} = 5;\ \overline{Y} = \frac{\Sigma Y}{N} = \frac{100}{25} = 4$$

$$r = \frac{520 - 25(5)(4)}{\sqrt{650 - 25(5)^2}\,\sqrt{436 - 25(4)^2}} = \frac{520 - 500}{\sqrt{25 \times 36}} = \frac{20}{30} = 0.667$$

Example 9:

In a question on correlation, the value of r is .917 and its probable error is 0.034. What would be the value of N ?

Solution:

$$\text{P.E. } r = .6745 \times \frac{1 - r^2}{\sqrt{N}}$$

We are given P.E. = .034, r = .917

Substituting the values $.034 = .6745 \times \dfrac{1 - (.917)^2}{\sqrt{N}}$

or $$.034 = \frac{.6745 \times .159}{\sqrt{N}}$$

or $.034\sqrt{N} = .1072$

$N = (3.153)^2 = 9.94$ or 10

Thus there were 10 pairs of observations.

Example 10:

The coefficient of rank correlation of the marks obtained by 10 students in statistics and accountancy was found to be 0.2. It was later discovered that the difference in ranks in the two subjects obtained by one of the students was wrongly taken as 9 instead of 7. Find the correct coefficient of rank correlation.

Solution:

$$R = 1 - \frac{6\Sigma D^2}{N^3 - N}$$

Here $R = 0.2,\ N = 10$

$$0.2 = 1 - \frac{6\Sigma D^2}{10^3 - 10}$$

$$0.2 - 1.0 = -\frac{6\Sigma D^2}{990} \quad -0.8 = -6\ \Sigma D^2 = 792 \text{ or } \Sigma D^2 = 132$$

Correct $\Sigma D^2 = 132 - (9)^2 + (7)^2 = 132 - 81 + 49 = 100$

Correct $R = 1 - \dfrac{6 \text{ Correct } \Sigma D^2}{10^3 - 10} = 1 - \dfrac{6 \times 100}{990} = 1 - 0.606 = 0.394$

Example 11:

Calculate correlation coefficient from the following results :

$n = 100,\ \Sigma X = 140,\ \Sigma Y = 150$

$= \Sigma (X - 10)^2 = 180,\ \Sigma (Y - 15)^2 = 215$

$= \Sigma (X - 10)^2\ (Y - 15)^2 = 60$

Solution:

For calculating correlation coefficient we need the values of $\Sigma X^2\ \Sigma Y^2\ \Sigma XY$ which we can determine from the values given.

$\Sigma (X - 10)^2 = \Sigma (X^2 + 100 - 2 \times 10 \times X)$

$= \Sigma X^2 + \Sigma 100 - 20\ \Sigma X$

$= \Sigma X^2 + 10 \times 100 - 20 \times 140$ (Since $\Sigma A = NA$)

$= \Sigma X^2 + 1000 - 2800$

$\Sigma X^2 = 180 - 1000 + 2800$ (Since $\Sigma (X - 10)^2 = 180$)

or $\quad \Sigma X^2 = 1980$

$$\Sigma (Y - 15)^2 = \Sigma (Y^2 + 225 - 2 \times 15 \times Y)$$

$$= \Sigma Y^2 + 225 - 30 Y$$

$$= \Sigma Y^2 + 225 \times 10 - 30 \times 150$$

$$= \Sigma Y^2 + 2250 - 4500$$

$$\Sigma Y^2 = 215 - 2250 + 2400 \qquad \text{(Since } \Sigma (Y - 15)^2 = 2215\text{)}$$

or $\quad \Sigma Y^2 = 2465 \qquad \Sigma (X - 10)(Y - 15) = 60$

or $\Sigma (XY - 15 \Sigma X - 10 \Sigma Y + 150) \times 10 = 60$

$$\Sigma XY - 15 \Sigma Y + 150 \times 10 = 60$$

$$\Sigma XY - 15 \times 140 - 10 \times 150 + 1500 = 60$$

$$\Sigma XY = 60 + 2100 + 1500 - 1500 = 2160$$

$$r = \frac{N \Sigma XY - \Sigma X \Sigma Y}{\sqrt{N \Sigma X^2 - (\Sigma X)^2}\sqrt{N \Sigma Y^2 - (\Sigma Y)^2}}$$

$$= \frac{10 \times 2160 - 140 \times 150}{\sqrt{10 \times 1980 - (140)^2}\sqrt{10 \times 2465 - (150)^2}}$$

$$= \frac{21600 - 21000}{\sqrt{19800 - 19600}\sqrt{N \Sigma Y^2 - (\Sigma Y)^2}}$$

$$= \frac{600}{\sqrt{200 \times 2150}} = \frac{600}{655.74} = +0.915$$

Example 12:

You are given the following information relating to a frequency distribution comprising 10 observations :

$$\overline{X} = 5.5, \ \overline{Y} = 4.0, \ \Sigma X^2 = 385, \ \Sigma Y^2 = 192$$

$\Sigma (X + Y)^2 = 947$. *Find* r_{xy}.

Solution:

$$\Sigma (X + Y)^2 = 947$$

$$\Sigma X^2 + \Sigma Y^2 + 2 \Sigma XY = 947$$

$$385 + 192 + 2 \Sigma XY = 947$$

$$2 \Sigma XY = 947 - 577 = 370$$

$$\Sigma XY = 185$$

$$\sigma_x = \sqrt{\frac{\Sigma X^2}{N} - (\overline{X})^2}$$

$$= \sqrt{\frac{385}{10} - (5.5)^2} = \sqrt{8.25} = 2.87$$

$$\sigma_y = \sqrt{\frac{\Sigma Y^2}{N} - (\bar{Y})^2}$$

$$= \sqrt{\frac{120}{10} - (4)^2} = \sqrt{3.2} = 1.79$$

$$r_{xy} = \frac{\frac{\Sigma XY}{N} - \overline{XY}}{\sigma_x \sigma_y} = \frac{\frac{185}{10} - (5.5)(4)}{2.87 \times 1.79}$$

$$= \frac{18.5 - 22}{5.14} = \frac{-3.5}{5.14} = 0.681.$$

Example 14:

In two sets of variables X and Y with 50 observations each, the following data were observed.

$$\bar{x} = 10,\ S.D.\ of\ X = 3,\ \bar{y} = 6,\ S.D.\ of\ Y = 2.$$

Coefficient of correlation between X and Y is 0.3. However, on subsequent verification it was found that one value of X (= 10) and one value of Y (= 6) were inaccurate and hence weeded out. With remaining 49 pairs of values, how is the original value of correlation coefficient affected?

Solution:

$$r = \frac{\frac{1}{N}(X-\bar{x})(Y-\bar{y})}{\sigma_x \sigma_y} \quad \text{or} \quad N \Sigma \sigma_x \sigma_y = (X-\bar{x})(Y-\bar{y})$$

$$\Sigma (X-\bar{x})(Y-\bar{y}) = 50 \times .3 \times 3 \times 2 = 90$$

$$\Sigma X Y = \bar{x} \Sigma y - \bar{y} \Sigma X + N \bar{x} \bar{y} = 90$$

$$\Sigma X Y = 90 + N \bar{x} \bar{y} \ [\because \Sigma Y = N \bar{y}, \Sigma X = N \bar{x}]$$

$$= 90 + (50 \times 10 \times 6)\ 3090$$

Correct $\Sigma X Y = 3090 - (10 \times 6) = 3030$

Correct $\Sigma X = (50 \times 10) - 10 = 490$

Correct $\Sigma Y = (50 \times 6) - 6 = 294$

$$\sigma_x^2 = \frac{\Sigma X^2}{N} - (\bar{x})^2$$

or $$9 = \frac{\Sigma X^2}{50} - (10)^2$$

$$450 - \Sigma X^2 - 5000$$

or $\Sigma X^2 = 5450$

Correct $\Sigma X^2 = 5450 - (10)^2 = 5350$

$$\sigma_y = \frac{\Sigma Y^2}{N} - (\bar{y})^2 \text{ or } 4 = \frac{\Sigma Y^2}{50} - (6)^2$$

$$200 - \Sigma Y^2 - 36 \times 50 \quad \text{or} \quad \Sigma Y^2 = 200$$

Correct $\Sigma Y^{21} = 200 - (6)^2 - 1964$

The correlation coefficient of remaining 49 pairs of values after weeding out the incorrect value shall be calculated as follows :

$$r = \frac{N \Sigma XY - \Sigma X \Sigma Y}{\sqrt{N \Sigma X^2 - (\Sigma X)^2} \sqrt{N \Sigma Y^2 - (\Sigma Y)^2}}$$

$$= \frac{49 \times 3030 - 490 \times 294}{\sqrt{49 \times 5350 - (490)^2} \sqrt{49 \times 1964 - (294)^2}}$$

$$= \frac{148470 - 144050}{\sqrt{262150 - 240100} \sqrt{96236 - 86436}}$$

$$= \frac{4410}{\sqrt{22050 \times 9800}} = 0.3$$

Example 14:

Coefficient of correlation between X and Y for 20 items is 0.3; mean of X is 15 and Y 20, standard deviations are 4 and 5 respectively. At the time of calculation one item 27 has wrongly been taken as 17 in case of X series and 35 instead of 30 in case of Y series. Find the correct coefficient of correlation.

Solution:

$$\bar{x} = \frac{\Sigma X}{N} \quad \text{or} \quad \Sigma X = N \bar{x}\text{; Here } N = 20, \ \bar{x} = 15$$

$\therefore$ $\Sigma X = 20 \times 15 = 300$

Correct $\Sigma X = 300 - 17 + 27 = 310$

Correct $$\bar{x} = \frac{310}{20} = 15.5$$

$$\bar{y} = \frac{\Sigma Y}{N} \text{ or } \Sigma Y = N\bar{y}$$

$$N = 20,\ \bar{y} = 20 \text{ S Y} = 20 \times 20 = 400$$

Correct $\text{S Y} = 400 - 35 + 30 = 395$

Correct $\bar{y} = \frac{395}{20} = 19.75$

Similarly, find correct standard deviation also

$$\sigma_x = \sqrt{\frac{\Sigma X^2}{N} - (\bar{x})^2}$$

$$4 = \sqrt{\frac{\Sigma X^2}{20} - (15)^2}, \text{ Squaring}$$

$$16 = \frac{\Sigma X^2}{20} = 225$$

$$320 - \Sigma X^2 - 4500 \text{ or } \Sigma X^2 = 4500 + 320 = 4820$$

Correct $\Sigma X^2 = 4820 - (17)^2 + (27)^2 = 5260$

Correct $\sigma_x = \sqrt{\frac{\text{Correct } \Sigma X^2}{N} - (\text{Correct } \bar{x})^2}$

$$= \sqrt{\frac{5260}{20} - (15.5)^2} = \sqrt{263 - 240.25} = 4.77$$

$$\sigma_x = \sqrt{\frac{\Sigma Y^2}{N} - (\bar{y})^2}$$

or $5 = \sqrt{\frac{\Sigma Y^2}{N} - (20)^2}$, Squaring

$$25 = \frac{\Sigma Y^2}{20} - 400$$

or $\Sigma Y^2 = 500 + 8000 = 8500$

Correct $\Sigma Y^2 = 8500 - (35)^2 + (30)^2 = 8175$

Correct $\sigma_y = \sqrt{\frac{8175}{20} - (19.75)^2} = \sqrt{408.75 - 390.68} = 4.$

Calculate S x y with uncorrected figures

$$r = \frac{\Sigma xy}{N\sigma_x \sigma_y}$$

Substituting the values, we get

$$.3 = \frac{\Sigma xy}{20 \times 4 \times 5} \quad \text{or} \quad \Sigma xy = 400 \times .3 = 120$$

Correct Σxy = Incorrect Σxy – (wrong item of X – Mean of X)

(wrong item of Y – Mean of Y)

+ (Correct item of X – Correct Mean of X)

(Correct item of Y – Correct Mean of Y)

$= 120 - (17 - 15)(35 - 20) + (27 - 15.5)(30 + 19.75)$

$= 120 - (2 \times 15) + (11.5 \times 10.25) = 120 - 30 + 117.875 = 207.875$

$$r = \frac{\Sigma xy}{N\sigma_x \sigma_y} = \frac{207.875}{20 \times 4.77 \times 4.323} = \frac{207.875}{412.414} = +0.504$$

Example 15:

Given $r - 0.8$, $S\,x\,y = 60$, $s_y = 2.5$ and $\Sigma x^2 = 90$, find the number of items, x and y are deviations from arithmetic mean.

Solution:

We are given

$$\Sigma x^2 = 90,\ r = 0.8,\ a_y = 2.5,\ \Sigma xy = 60$$

We have to find the number of observations, i.e., n

$$\sigma_x^2 = \frac{1}{n}\Sigma(X - \bar{x})^2 = \frac{1}{n}\Sigma x^2 = \frac{90}{n}$$

$$r = \frac{\Sigma(X - \bar{x})(Y - \bar{y})}{n\sigma_x \sigma_y} = \frac{\Sigma xy}{n\sigma_x \sigma_y}. \text{ Squaring}$$

$$r^2 = \frac{(\Sigma xy)^2}{n^2 \sigma_x{}^2 \sigma_y{}^2}$$

$$(0.8)^2 = \frac{(60)^2}{n^2 \times \left(\frac{90}{n}\right) \times 6.25} = \frac{3600}{90n \times (6.25)}$$

$$.64 = \frac{3600}{562.5n}$$

$.64 \times 562n = 3600$ or $360\,n = 3600$ or $n = 10$.

Example 16:

Compute Karl Pearson's coefficient of correlation between the capita National Income and per capita Consumer Expenditure from the data given below :

Years	*Per capita National Income*	*Per capita Consumer Expenditure*
1988	*249*	*237*
1989	*251*	*238*
1990	*248*	*236*
1991	*252*	*240*
1992	*258*	*245*
1993	*269*	*255*
1994	*271*	*254*
1995	*272*	*252*
1996	*280*	*258*
1997	*275*	*251*

Also calculate probable error.

Solution:

Since the actual means of X and Y variable are not whole number, we should only the assumed mean method of finding correlation.

Calculation of Karl Pearson's Coefficient

X	**(X – A)** d_x	d_x^2	**Y**	**(Y – A)** d_y	d_y^2	$d_x d_y$
249	– 11	121	237	– 3	9	33
251	– 9	81	238	– 2	4	+ 18
248	– 12	144	236	– 4	16	+ 48
252	– 8	64	240	0	0	0
258	– 2	4	245	+ 5	25	– 10
269	+ 9	81	255	+ 15	225	+ 135
271	+ 11	121	254	+ 14	196	+ 154
272	+ 12	144	252	+ 12	144	+ 144
280	+ 20	400	258	+ 18	324	+ 360
275	+ 15	225	251	+ 11	121	+ 165
N = 2625	$\Sigma d_x = 25$	$\Sigma d_x^2 = 135$	$\Sigma Y =$ 2466	$\Sigma d_y = 66$	$\Sigma d_y^2 =$ 1064	$\Sigma d_x d_y =$ 1047

$$r = \frac{\Sigma d_x d_y \quad \frac{(\Sigma d_x)(\Sigma d_y)}{N}}{\sqrt{\Sigma d_x^2 - \frac{(\Sigma d_x)^2}{n}} - \sqrt{\frac{\Sigma d_y^2 - (\Sigma d_y)^2}{n}}}$$

Taking deviations of X series from 260 and that of Y series from 240.

$\Sigma d_x d_y = 1047$, $\Sigma d_x = 25$, $\Sigma d_y = 66$

$\Sigma d_x^2 = 1385$, $\Sigma d_y^2 = 1064$, $N = 10$

Substituting the values

$$r = \frac{1047 - 25 \times \frac{66}{10}}{\sqrt{1385 - \frac{25^2}{10}}\sqrt{1064 - \frac{(66)^2}{10}}}$$

$$= \frac{1047 - 165}{\sqrt{1385 - 62.5}\sqrt{1064 - 435.6}}$$

$$= \frac{882}{36.367 \times 25.07} = \frac{882}{911.65} = 0.967$$

$$P.E_r = 0.6745 \frac{1 - r^2}{\sqrt{N}}$$

$$= 0.6745 \frac{1 - (.96)^2}{\sqrt{10}} = \frac{0.6745 \times .065}{3.162} = 0.014$$

Example 17:

From the following table calculate the coefficient of correlation by Karl Pearson's method :

X :	*6*	*2*	*10*	*4*	*8*
Y :	*9*	*11*	*?*	*8*	*7*

Arithmetic means of X and Y series are 6 and 8 respectively.

Solution:

From the mean value of Y we can find out value.

$$Y = \frac{\Sigma Y}{N}; \; 8 = \frac{\Sigma Y}{5} \text{ or } \Sigma Y = 40.$$

The given values are 9 + 11 + 8 + 7 = 35.

Hence the missing value is 40 – 35 = 5

Calculation of Correlation Coefficient

X	(X – 6) x	x^2	Y	$(Y - \bar{y})$ y	y^2	xy
6	0	0	9	+ 1	1	0
2	– 4	16	11	+ 3	9	– 12
10	+ 4	16	5	– 3	9	– 12
4	– 2	4	8	0	0	0
8	+ 2	4	7	– 1	1	– 2
$\Sigma X = 30$	$\Sigma x = 0$	$\Sigma x^2 = 40$	$\Sigma Y = 40$	$\Sigma Y = 0$	$\Sigma Y^2 = 20$	$\Sigma xy = -26$

$$r = \frac{\Sigma xy}{\sqrt{\Sigma x^2 \times \Sigma y^2}} = \frac{-26}{\sqrt{40 \times 20}}$$

$$= \frac{-26}{28.28} = -0.919$$

Example 18:

Calculate Karl Pearson's coefficient of correlation for the data given below, taking 66 and 63 as assumed means of X and Y respectively.

Height of Husbands X (in inches) :	*60*	*62*	*64*	*66*	*68*	*70*	*72*
Height of Wives Y :	*61*	*63*	*63*	*63*	*64*	*65*	*67*

Solution:

Calculation of Coefficient of Correlation

X	(X – 66) d_x	d_x^2	Y	(Y – 63) d_y	d_y^2	$d_x d_y$
60	– 6	36	61	– 2	4	12
62	– 4	16	63	0	0	0
64	– 2	4	63	0	0	0
66	0	0	63	0	0	0
68	+ 2	4	64	+ 1	1	2
70	+ 4	16	65	+ 2	4	8
72	+ 6	36	67	+ 4	16	24
	$\Sigma d_x = 0$	$\Sigma d_x^2 = 112$		$\Sigma d_y = 5$	$\Sigma d_y^2 = 25$	$\Sigma d_x d_y = 46$

$$r = \frac{N\Sigma d_x d_y - \Sigma d_x \Sigma d_y}{\sqrt{N\Sigma d_x^2 - (\Sigma d_x)^2}\sqrt{N\Sigma d_x^2 - (\Sigma d_y)^2}}$$

$$= \frac{7 \times 46 - 0 \times 5}{\sqrt{1 \times 112 - (0)^2}\sqrt{7 \times 25 - (5)^2}}$$

$$= \frac{322}{\sqrt{784 \times 150}} = \frac{322}{342.93} = 0.939$$

Example 19:

Calculate Karl Pearson's coefficient of correlation taking deviations from actual mean 52 (X series) and (Y series)

X Series :	*44*	*46*	*46*	*48*	*52*	*54*	*?*	*56*	*60*	*60*
Y Series :	*36*	*40*	*42*	*40*	*?*	*44*	*46*	*48*	*50*	*52*

Solution:

We have one mixing value in series X and one in Y. Since mean of X is 52

$\Sigma X = N\bar{X}$ i.e., $10 \times 52 = 520$

Total of 9 values of series X = 466

Missing value = 54

$\Sigma Y = N\bar{y}$ i.e., $10 \times 44 = 440$

Total of 9 values of series Y = 398

Missing value = 42

Calculation of Correlation Coefficient

X	(X – 52) x	x^2	Y	(Y – 44) y	y^2	xy
44	– 8	64	36	– 8	64	64
46	– 6	36	40	– 4	16	24
46	– 6	36	42	– 2	4	12
48	– 4	16	40	– 4	16	16
52	0	0	42	– 2	4	0
54	+ 2	4	44	0	0	0
54	+ 2	4	46	+ 2	4	4
56	+ 4	16	48	+ 4	16	16
60	+ 8	64	50	+ 6	36	48
60	+ 8	64	52	+ 8	64	64
$\Sigma X = 520$	$\Sigma x = 0$	$\Sigma x^2 = 304$	$\Sigma Y = 440$	$\Sigma y = 0$	$\Sigma y^2 = 224$	$\Sigma xy = 248$

$$r = \frac{\Sigma xy}{\sqrt{\Sigma x^2 \times \Sigma y^2}} = \frac{248}{\sqrt{304 \times 224}} = \frac{248}{260.952} = 0.92.$$

Example 20:

Calculate Karl Pearson's coefficient of correlation from the following data :

Y \ *X*	*200–300*	*300–400*	*400–500*	*500–600*	*600–700*
10–15	—	—	—	*3*	*7*
15–20	—	*4*	*9*	*4*	*3*
20–25	*7*	*6*	*12*	*5*	—
25–30	*3*	*10*	*19*	*8*	—

Solution:

Computation of Correlation Coefficient

Y \ X	d_y \ d_x	200–300 250 − 2	300–400 300 − 1	400–500 450 0	500–600 550 + 1	600–700 650 + 2	f	$f\,d_y$	$f\,d_y^2$	$f\,d_x d_y$
10–15	− 1	—	—	—	− 3 3	+ 14 7	10	− 10	10	− 17
15–20	0		0 4	0 9	0 4	0 3	20	0	0	0
20–25	1	− 14 7	6 6	0 12	5 5	—	30	30	30	− 15
25–30	2	12 3	− 20 10	0 19	16 8	—	40	480	160	− 16
	f	10	20	40	20	10	N = 100	Σfd_y = 100	Σfd_y^2 = 200	Σfd_xd_y = − 48
	$f\,d_x$	− 20	− 20	0	+ 20	+ 20	Σfd_x = 0			
	$f\,d_x^2$	40	20	0	20	40	Σfd_x^2 = 120			
	$f\,d_xd_y$	− 26	− 26	0	18	− 14	Σfd_xd_y = − 48			

$$r = \frac{N\Sigma f d_x d_y - \Sigma f d_x \Sigma f d_y}{\sqrt{N\Sigma f d_x^2 - (\Sigma f d_x)^2}\sqrt{N\Sigma f d_y^2 - (\Sigma f d_y)^2}}$$

$\Sigma f d_x d_y = -48,$

$\Sigma f d_x = 0, \Sigma f d_y = 100,$

$\Sigma f d_x^2 = 120, \Sigma f d_y^2 = 200, \quad N = 100$

$$r = \frac{100(-48) - 0 \times 100}{\sqrt{100(120) - (0)^2}\sqrt{100(200) - (100)^2}}$$

$$= \frac{-4800}{\sqrt{12000}\sqrt{10000}}$$

$$= \frac{-4800}{10954} = -0.438.$$

Example 21:

In a question on correlation, the value of r is .917 and its probable error is 0.034. What would be the value of N ?

Solution:

$$\text{P.E. } r = .6745 \times \frac{1-r^2}{\sqrt{N}}$$

We are given P.E. = .034, r = .917

Substituting the values $.034 = .6745 \times \dfrac{1-(.917)^2}{\sqrt{N}}$

or $$.034 = \frac{.6745 \times .159}{\sqrt{N}}$$

or $$.034\sqrt{N} = .1072$$

$N = (3.153)^2 = 9.94$ or 10

Thus there were 10 pairs of observations.

Example 22:

Calculate Pearson's coefficient of correlation from the following data using 44 and 26 as the origin of X and Y respectively :

X :	43	44	46	40	44	42	45	42	38	40	42	57
Y :	29	31	19	18	19	27	27	29	41	30	26	10

Solution:

Calculation of Pearson's Coefficient of Correlation

X	(X – 44) d_x	d_x^2	Y	(Y – 26) d_y	d_y^2	$d_x d_y$
43	– 1	1	29	+ 3	9	– 3
44	0	0	31	+ 5	25	0
46	+ 2	4	19	– 7	49	– 18
40	– 4	16	18	– 8	64	+ 32
44	0	0	19	– 7	49	– 14
42	– 2	4	27	+ 1	1	– 2
45	+ 1	1	27	+ 1	1	+ 1
42	– 2	4	29	+ 3	9	– 6
38	– 6	36	41	+ 15	225	– 90
40	– 4	16	30	+ 4	16	– 16
42	– 2	4	26	0	0	0
57	+ 13	169	10	– 16	256	– 208
	$\Sigma d_x = -5$	$\Sigma d_x^2 = 225$		$\Sigma d_y = -6$	$\Sigma d_y^2 = 704$	$\Sigma d_x d_y = 306$

$$r = \frac{N \Sigma d_x d_y - \Sigma d_x d_y}{\sqrt{N \Sigma d_x^2 - (\Sigma dx)^2}\ \sqrt{N \Sigma d_y^2 - (\Sigma d_y)^2}}$$

$$= \frac{12(-306) - (-5)(-6)}{\sqrt{12 \times 225 - (-5)^2}\ \sqrt{12 \times 704 - (-6)^2}}$$

$$= \frac{-3672 - 30}{\sqrt{3080 - 25}\ \sqrt{8448 - 36}} = \frac{-3702}{\sqrt{3035}\ \sqrt{8412}}$$

$$= \frac{-3702}{55.09 \times 91.72} = \frac{-3702}{5052.85} = -0.733.$$

Example 23:

The coefficient of rank correlation of the marks obtained by 10 students in statistics and accountancy was found to be 0.2. It was later discovered that the difference in ranks in the two subjects obtained by one of the students was wrongly taken as 9 instead of 7. Find the correct coefficient of rank correlation.

Solution:

$$R = 1 - \frac{6\Sigma D^2}{N^3 - N}$$

Here $R = 0.2,\ N = 10$

$$0.2 = 1 - \frac{6\Sigma D^2}{10^3 - 10}$$

$$0.2 - 1.0 = -\frac{6\Sigma D^2}{990} \quad -0.8 = -6\ \Sigma D^2 = 792 \quad \text{or} \quad \Sigma D^2 = 132$$

Correct $\Sigma D^2 = 132 - (9)^2 + (7)^2 = 132 - 81 + 49 = 100$

Correct $R = 1 - \frac{6 \text{ Correct } \Sigma D^2}{10^3 - 10} = 1 - \frac{6 \times 100}{990} = 1 - 0.606 = 0.394$

LIST OF FORMULAE

1. Karl Pearson's Correlation Coefficient (when deviations are taken from actual means)

$$r = \frac{\Sigma xy}{N\sigma_x \sigma_y} \quad \text{or} \quad \frac{\Sigma xy}{\sqrt{\Sigma x^2 \times \Sigma y^2}}$$

where $x = (X - \overline{X})$;

$y = (Y - \overline{Y})$.

2. When deviations are taken from assumed mean

$$r = \frac{N\Sigma d_x d_y - \Sigma d_x \times \Sigma d_y}{\sqrt{N\Sigma d_x^2 - (\Sigma d_x)^2}\ \sqrt{N\Sigma d_y^2 - (\Sigma d_y)^2}}$$

where $d_x = (X - A)$ and $d_y = (Y - A)$

3. In a bivariate frequency distribution :

$$r = \frac{N\ \Sigma\ fd_x d_y - \Sigma fd_x \Sigma fd_y}{\sqrt{N\Sigma fd_x^2 - (\Sigma fd_x)^2}\ \sqrt{N\Sigma fd_y^2 - (\Sigma fd_y)^2}}$$

4. When we deal with actual values of X and Y :

$$\frac{\Sigma\ XY - \Sigma X \Sigma Y}{\sqrt{N\Sigma X^2 - (\Sigma X)^2}\ \sqrt{N\Sigma Y^2 - (\Sigma Y)^2}}$$

5. Spearman's Rank Correlation Coefficient

$$R = 1 - \frac{6\Sigma D^2}{N^3 - N}$$

In case ranks are repeated

$$R = 1 - \frac{6\left(\Sigma D^2 + \frac{1}{12}(m^2 - m) + \frac{1}{12}(m^3 - m)\right)}{N^3 - N}$$

6. Concurrent Deviation Method

$$r_C = \pm\sqrt{\left(\frac{2C - n}{n}\right)}$$

$$S.E._r = \frac{1 - r^2}{\sqrt{N}};\ P.E._r = 0.6745\,\frac{1 - r^2}{\sqrt{N}}$$

Coefficient of determination = r^2.

EXERCISES

1. Find out rank coefficient of correlation from the following :

A :	115	109	112	87	98	98	120	100	98	118
B :	75	73	85	70	76	65	82	73	68	80

2. From the following data, find correlation coefficient between age and blindness :

Age (years)	*No. of Persons (000)*	*No. of Blinds*
0 –10	100	45
10 – 20	50	40
20 – 30	40	40
30 – 40	32	40
40 –50	24	36
50 – 60	11	22
60 – 70	6	18
70 – 80	8	15

3. Determine the correlation from the following value of X and Y:

X :	114	106	96	116	120	118	94	99	106	108	117
Y :	66	69	83	64	60	63	82	80	71	70	65

4. From the following data, calculate the coefficient of correlation by Karl Pearson's method :

X :	6	2	10	4	8
Y :	9	11	9	8	7

Arithmetic mean of X and Y series are 6 and 8 respectively.

5. Seven students obtained the following percentage of marks in the college test (X) and the final examination (Y). Find the coefficient of correlation between these variables :

X :	50	62	72	25	20	60	60
Y :	43	65	74	33	25	55	66

6. Correlation between two variables has a value of 0.9 and a correlation between other two variables is 0.3. Can we infer that the first correlation is twice as strong as the second? Give reasons.

7. Compute the coefficients of correlation between the number of vehicles and the sale of tyres from the data given below :

No. of Vehicles ('000):	5.4	5.6	5.8	6.2	6.4	5.5
Sales of tyres ('000) :	11.8	12.0	14.5	16.3	13.8	12.2

8. The scores of 8 students in an examination in Mathematics and Statistics are given below :

Roll No.:	1	2	3	4	5	6	7	8
Marks in Mathematics :	70	48	58	55	54	50	60	52
Marks in Statistics:	62	47	53	60	55	68	51	48

Find : (i) Product moment correlation coefficient, and (ii) Rank correlation coefficient and [(i) r = 0.246 (ii) R = 0.286].

9. From the following data, calculate the coefficient of correlation between age of students and their playing habit :

Age :	14	15	16	17	18
No. of Students :	300	200	150	120	100
Regular Players :	225	130	90	48	35

10. Find the correlation coefficient between age and playing habits of the following students :

Age :	15	16	17	18	19	20
No. of students :	250	200	150	120	100	80
Regular players :	200	150	90	48	30	12

11. Distinguish giving suitable examples between :
    (i) Positive and negative correlation.
    (ii) Linear and non-linear correlation.

(iii) Simple, partial and multiple correlation.

12. (a) Explain the concept of correlation. Describe the method of measuring correlation from a bivariate table.

    (b) Prove that the correlation coefficient is unaffected by the change of origin and scale.

13. What is correlation? Distinguish between positive and negative correlation. Explain the significance of coefficient of correlation.

14. (a) What assumptions must be made in order to use Pearsonian product moment correlation coefficient?

    (b) Show that for two independent variates, correlation coefficient is zero.

15. Ten students obtained the following marks in Statistics (X) and Accountancy (Y) :

Student:	A	B	C	D	E	F	G	H	I	J
X:	92	89	86	87	83	71	77	63	53	50
Y:	86	83	77	91	68	52	85	82	57	57

Find the rank correlation coefficient.

16. The following data relate to sales and expenses of 10 firms. Calculate its coefficient of correlation by direct method :

Firm :	A	B	C	D	E	F	G	H	I	J
Sales (in Rs. '000):	50	50	55	60	65	65	65	60	60	50
Expenses (Rs.'000):	11	13	14	16	16	15	15	14	13	13

17. Compute Karl Pearson's coefficient of correlation between per capita national income and per capita consumer expenditure from the data given below :

Years	*Per Capita N.I. Expenditure*	*Per Capita Consumer*
1985	249	237
1986	251	238
1987	248	236
1988	252	240
1989	258	245
1990	269	255

1991	271	254
1992	272	252
1993	280	268
1994	290	271

18. Calculate coefficient of correlation between X and Y from the following series :

	X	*Y*
No. of observations :	10	10
Mean :	20	16
Sum of squares of deviations from the mean :	30	108
Total of products of deviations of X and Y series from their respective means :	35	

19. A group of eight students got the following percentage of marks in a test in Statistics and Accountancy :

Roll No. :	11	12	13	14	15	16	17	18
% of marks (Statistics) :	50	60	65	70	75	40	70	80
% of marks (Accountancy) :	80 71	60	75	90	82	70	50	

Compute correlation coefficient.

20. From the data given below, compute Karl Pearson's coefficient of correlation :

	X-Series	*Y-Series*
Number of Items :	15	15
Arithmetic Mean :	25	18
Square of deviation from Arithmetic Mean :	136	138
Summation of Products of deviations from Arithmetic Means of X and Y :		122

21. From the following data of the marks obtained by 8 students in the accountancy and statistics papers, compute the rank coefficient of correlation :

Marks in Accountancy :	15	20	28	12	40	60	20	80
Marks in Statistics :	40	30	50	30	20	10	30	60

22. Calculate Karl Pearson's coefficient of correlation having deviations from actual means 52 and 44 of the following data :

X :	44	46	46	48	52	54	X	56	60	60
Y :	36	40	42	40	Y	44	46	48	50	52

23. Explain how covariance of X and Y is related to the coefficient of simple correlation between X and Y.

24. (a) Does the correlation coefficient imply 'causation' between two variables?

    (b) What is Rank Correlation? State merits and demerits of Spearman's Rank Correlation method.

25. (a) Prove that Karl Pearson's coefficient of correlation always lies between ± 1.

    (b) What are the limits of the value of 'r'? What do positive, negative and zero values of 'r' indicate?

26. (a) What is meant by correlation? What is the significance of positive and negative correlation?

    (b) Distinguish between simple, multiple and partial correlation.

27. Define Karl Pearson's coefficient of correlation. What is it intended to measure? How would you interpret the sign and magnitude of a correlation coefficient?

28. (a) Give an example each of "spurious correlation" and "nonsense or chance correlation".

    (b) What is correlation? Explain how will you use the following methods of determining correlation:

    (i) Graph (ii) Correlation table (iii) Karl Pearson's coefficient of correlation.

29. (a) Define Karl Pearson's coefficient of correlation. Interpret r when r = 1, – 1 or 0.

    (b) A correlation coefficient, r = 0.8 indicates a relationship twice as close as r = 0.4, Comment.

30. (a) State the properties and significance of Pearson's correlation coefficient.

    (b) How is scatter diagram useful in the study of correlation?

    (c) What is correlation? How does it help in studying the correlation between two variables in respect both its nature and extent?

31. (a) Explain the assumption on which Karl Pearson's coefficient of correlation is based.

    (b) Define Karl Pearson's coefficient of correlation. What does it intend to measure? Show that it is not affected by the change of origin and scale.

32. (a) Explain what is meant by correlation between two variables.

What are the methods of finding the existence of correlation? How can it be measured?

(b) Define Karl Pearson's coefficient of correlation. What does it intend to measure? Show that it is not affected by the change of origin and scale.

33. (a) Explain what is meant by correlation between two variables. What are the methods of finding the existence of correlation? How can it be measured?

(b) Define rank correlation coefficient. How is it determined and when it preferred to Karl Pearson's coefficient?

34. (a) What is scatter diagram? How does it help in studying the correlation between two variables in respect of both of its direction and degree?

(b) What are the methods of calculating coefficient of correlation?

35. Comment on the following:

(i) "Independence of Y and X implies zero correlation but not vice versa."

(ii) "A correlation coefficient of 0.6 indicates a relationship twice as close are r = 0.3."

(iii) "Even a high degree of correlation does not mean that a relationship of cause and effect exists between the two correlated variables." Why?

(iv) "Rank correlation coefficient is especially useful under certain situation."

(v) What are the main types of correlations?

36. (a) Explain the meaning and significance of the concept of correlation. Does it always signify cause and effect relationship between two variables? How is the coefficient of correlation interpreted?

(b) Explain the meaning of correlation between two variables. State the methods available for finding the correlation.

37. (a) What is correlation? Give the properties of Karl Pearson's coefficient of correlation.

(b) What is correlation? Clearly explain, with suitable illustrations, its role in dealing with business problems.

38. (a) Prove that the correlation coefficient is unaffected by the change of origin or scale.

(b) Define rank correlation coefficient. When is it preferred to Karl Pearsons's coefficient of correlation?

39. Explain the maning of correlation. State the extreme values of the coefficient of correlation and interpret them.

40. State the properties of Pearson's coefficient of correlation. How do you interpret a calculated value of r? Explain the term "probable error of r."

41. (a) What is correlation? Explain the significance of positive, negative or zero coefficient of correlation.

    (b) Explain the term correlation and give methods to measure the same.

42. (a) Define Karl Pearson's coefficient of correlation. Show that it is independent of change of scale and origin.

    (b) Prove that two independent variables are uncorrelated. By giving an example, show that the converse is not true. Explain the reason?

43. (a) What is Spearman's rank correlation coefficient? How does it difference from Karl Pearson's coefficient of correlation.

    (b) Differentiate between the coefficient of correlation and coefficient of determination.

44. (a) Define Rank Correlation. Write down Spearman's formula for rank correlation coefficient. State the advantages of Spearman's rank correlation over Karl Pearson's correlation coefficient.

    (b) Define coefficient of correlation between two variables. What is the range of the coefficient? Interpret the results: $r = 0$. $r = \pm 1$.

45. (a) What is rank correlation? State the merits and demerits of Spearman's rank correlation method.

    (b) Define positive correlation and negative correlation giving examples. Define covariance between two variables.

46. (i) If the coefficient of correlation between the annual value of exports during the last ten years and the annual number of children born during the same period is + 0.9, what inference, if any, would you draw?

    (ii) If the coefficient between two variables x and y is positive, then the coefficient of correlation between – x and – y is also positive. Comment.

    (iii) If the correlation coefficient between two variables is zero, then the variables are not necessarily uncorrelated. Comment.

    (iv) Does a zero value of Karl Pearson's coefficient of correlation between twᴖ variables X and Y imply that X and Y are not related? Explain.

47. (i) Write down an expression for the Karl Pearson's coefficient of linear correlation. Why is it termed as the coefficient of linear correlation? Explain.

(ii) The correlation coefficient between two variables x and y is found to be 0.4. What is the correlation between 2x and (y)?

(iii) Interpret 'r' in the following cases :

when r = + 0.97, r = – 0.04, r = + 1, r = – 1.

48. Tick the correct answer :

(a) The coefficient of correlation :

(i) has to limits (ii) can be less than 1 (iii) can be more than 1 (iv) varies between ± 1 (v) none of these.

(b) The product moment correlation coefficient is obtained by the formula:

(i) r = RXY/xy

(ii) $r = \Sigma\ xy/N\ \sigma_x\ \sigma_y$

(iii) r – Σ XY/N sx

(iv) ΣxΣ y/σx σy (v) None of these.

(c) Probable error is :

(i) .06745 S.E. (ii) .6457 S.E. (iii) 6753 S.E. (iv) .7653 S.E. (v) .6547 S.E.

(d) Rank correlation coefficient is obtained by the formula :

(i) $R = 1 + 6\Sigma\ D^2/N^3 - N$ (ii) $R = 1 + 6\Sigma\ D/N^2 - N$ (iii) $R = 1 - 62\ D^2 - N^2$ (iv) $R = 1 - 6\Sigma D/N\ (N^3 - 1)$ (v) $R = 1 - \Sigma D^2/N^3 - N$

(e) Coefficient of determination is defined as

(i) r^3 (ii) $1 - R^2$ (iii) $1 + r^2$ (iv) r^2 (v) none of these.

(f) The limits of the population correlation are given by :

(i) r ± P.E. (ii) r ± 3 S.E. (iii) r ± 3 P.E. (iv) r ± 3 S.E. (v) none of these.

(g) If r is 0.8, coefficient of alternation shall be :

(i) 0.64 (ii) 0.4 (iii) 0.6 (iv) 0.8 (v) none of these.

(h) If sum of the product of deviations of X and Y series from their means, i.e. ΣXY is zero, the coefficients of correlation shall be :

(i) + 1 (ii) 0 (iii) – 1 (iv) none of these.

(i) Mention the correct answer :

The ranks according to two attributes in a sample are given below :

R_1 :	1	2	3	4	5
R_2 :	5	4	3	2	1

The rank correlation between them is

(i) 0 (ii) + 1 (iii) – 1 (iv) none of these.

(j) While drawing a scatter diagram if all points appear to form a straight the going downward from left to right, then it is inferred that there is ....

(i) Perfect positive correlation.

(ii) Simple positive correlation.

(iii) Perfect negative correlation.

(iv) No correlation.

[**Ans.** a. (iv), b. (ii) c. (i) d. (iii) e. (iv) f. (i), g. (iii) h. (ii), i. (iii) j. (iv)].

49. State the nature of correlation (positive, negative or no correlation) between :

(i) Sale of woolen garments and the day temperature.

(ii) Income and expenditure of household.

(iii) Height of student and the marks obtained by him.

(iv) Total investment and the rate of interest.

50. (a) From the data given below, find the coefficient of correlation between the driver's age and the number of accidents made by him :

Number of accidents	*Driver's age*				
	25 – 30	30 – 35	35 – 40	40 – 45	45 – 50
0	–	3	3	7	8
1	–	–	9	4	1
2	3	5	10	3	–
3	4	9	6	–	–
4	12	7	3	1	–

51. In order to find the correlation coefficient between two variables X and Y from 12 pairs of observations, the following calculations were made :

S X = 30, S Y = 5.S X^2 = 670, S Y^2 = 285, S XY = 334.

On subsequent verification it was found that the pair (X = 11, Y = 4) was copied wrongly, the correct value being X = 10, Y = 14. Find the correct value of r.

52. Two series X and Y 50 items each have standard deviation 4.5 and 3.5 respectively. If the summation of the products of deviations of the two series from their respective means be 420, find the coefficient of correlation between X and Y.

53. From the following table, find the correlation between age and playing habit :

Age (years)	*No. of Students*	*Regular players*
15 – 16	200	150
16 – 17	270	162
17 – 18	340	170
18 – 19	360	180
19 – 20	400	180
20 – 21	200	120

54. A computer, while calculating the correlation between two variables 5X and Y, obtained the following constants :

$N = 30, \Sigma X = 120, \Sigma X^2 = 600, \Sigma Y = 90, \Sigma Y^2 = 250, \Sigma XY = 356.$

It was, however, later discovered at the time of checking that it had copied down two pairs of observations as :

X	*Y*
8	10
10	7

While the correct values were :

X	*Y*
8	12
10	8

Obtain the correct value of the correlation coefficient between X and Y.

55 (a) Explain what is meant by coefficient of correlation between two variables. What are the different methods of finding correlation? Distinguish between Positive and Negative correlation.

(b) Define coefficient of correlation and mention its important properties.

56. (a) In a bivariate sample, the sum of squares of difference between the ranks of observed values of two variables is 231 and the

correlation coefficient between them is 0.4. Find the number of pairs.

(b) Explain the concepts of (i) Coefficient of determination, and (ii) Probable error. How are they helpful in interpreting correlation?

(c) Interpret the value of 0 and +1 for the product moment correlation coefficient.

57. Family income (X) and percentage expenditure on food (Y) in the case of hundred families gave the following frequency distribution:

% Expenditure on food	*Family income (Rs.)*				
	200–300	300–400	400–500	500–600	600–700
20 – 25	–	–	–	3	7
25 – 30	–	4	9	4	3
30 – 35	7	6	12	5	–
35 – 40	3	10	19	8	–

43. Calculate the coefficient of correlation between marks in Statistics and in Economics of 100 students who appeared in the B. Com. examination:

Marks in Economics	*Marks in Statistics*				
	0 – 10	10 – 20	20 –30	30 – 40	40 – 50
0 – 10	6	33	–	–	–
10 – 20	3	16	10	–	–
20 – 30	–	10	15	7	–
30 – 40	–	–	7	10	–
40 – 50	–	–	–	4	5

58. (a) What are the advantages of the rank correlation coefficient ?

(b) Distinguish between Positive and Negative correlation.

(c) What is meant by rank correlation?

(d) What is the use of scatter diagram?

(e) Explain rank correlation with an example.

59. (a) What is meant by the correlation between two variables?

(b) What is scatter diagram? What is its use?

(c) Explain what is meant by perfect correlation.

60. (a) Name the different measures of correlation and briefly discuss their uses.

(b) Compare Spearman's correlation coefficient with Karl Pearson's correlation coefficient.

(c) Define correlation. Discuss various types of correlations.

61. (a) What is 'Spurious' or 'non-sensical correlation'? Explain with an example.

(b) What is coefficient of Rank Correlation? Bring out its usefulness. How does it differ from coefficient of correlation?

62. Fill in the blanks :

(i) When the relationship is of a .... nature, the appropriate statistical tool for discovering and measuring the relationship and expressing the brief formula is known as ....

(ii) Positive correlation implies that, on an average as one variable is increasing, the other .... is and as one variable is decreasing, the other is also ....

(iii) The relationship between three or more variables is studied with the help of .... correlation.

(iv) If r is more than six times .... it is called significant.

(v) The probable error of correlation coefficient is 0.6745 of ....

(vi) If r = 0.3, r^2 will be ....

(vii) The coefficient of correlation is independent of change of .... and ....

(viii) The coefficient of correlation is under-root of two ....

[**Ans.** (i) quantitative, correlation (ii) increasing, decreasing (iii) multiple (iv) probable error (v) standard error (vi) 0.09 (vii) scale, origin (viii) regression coefficients].

63. The following data relate to the prices and supplies of a commodity during a period of eight years :

Price (Rs./kg) :	10	12	18	16	15	19	18	17
Supply (100 kg) :	30	35	45	44	42	48	47	46

Calculate the coefficient of correlation between the two series.

64. Seven methods of imparting business education were ranked by the MBA students of two universities as follows :

Methods of teaching :	I	II	III	IV	V	VI	VII
Rank of students of							
University A :	2	1	5	3	4	7	6

Calculate rank correlation coefficient.

65. Indicate whether the following statements are *True* or *False* :

    (i) There are no limits to the value of r. T/F

    (ii) If r is negative, both the variables are decreasing. T/F

    (iii) Correlation always signifies a cause and effect relationship between the variables. T/F

    (iv) The rank correlation coefficient was developed by Spearman. T/F

    (v) If the values of X variable are 1, 2, 3, 4, 5 and those of Y, 4, 6, 8, 10, 12, the Karl Pearson's and the Rank method would give the same answer. T/F

    (vi) Pearsonian coefficient is the best under all situations. T/F

    (vii) Before calculating and interpreting the value of r. Utmost care must be exercised to see what variables are being studied. T/F

    (viii) Nonsense or spurious correlation implies real relationship between the variables. T/F

    (ix) Negative correlation in two series means that, as value of one of the variables, decreases the value of the other variable would also decrease. T/F

    [**Ans.** (i) F (ii) F (iii) F (iv) T (v) T (vi) F (vii) T (viii) T (ix) F].

2

Regression Analysis

INTRODUCTION

The term 'regression' was first used by Sir Francis Galton in 1877while studying the relationship between the height of fathers and sons. The term is still used to describe that line drawn for a group of points to represent the trend present, but it no longer necessarily carries the original implication of "stepping back" that Galton intended. These days there is a growing tendency of the modern writers to use the term *estimating line instead of regression line* because the expression estimating line is more clarificatory in character.

Some most important definitions are given here:

1. The term 'regression analysis' refers to the methods by which estimates are made of the values of a variable from a knowledge of the values of one or more other variables and to the measurement of the errors involved in this estimation process." *–Morris Hamburg*
2. "Regression is the measure of the average relationship between two or more variables interms of the original units of the data."
3. "Regression analysis attempts to establish the '*nature of the relationship*' between variables–that is, to study the functional relationship between the variables and thereby provide a mechanism for prediction, or forecasting." *–Ya–Lum Chou*
4. "One of the most frequently used techniques in economics and business research, to find a relation between two or more variables that are related causally, is regression analysis." *–Taro Yamane*

The unknown values of one variable from known values of another variable. The variable which is used to predict the variable of interest is called the independent variable or explanatory variable and the variable we are trying to predict is called the dependent variable or "explained" variable. The independent variable is denoted by X and the dependent variable by

Y. The analysis used is called the simple linear regression analysis–simple because there is only one predictor or independent variable, and linear because of the assume linear relationship between the dependent and the independent variables. The term "linear" means than an equation of a straight line of the form Y = a + bX, where a and b are constants, is used to describe the average relationship that exists between the two variables.

It meant is simply that estimates of values of the dependent variable Y may be obtained for given values of the independent variable X for a mathematical function involving X and Y. In that sense, the values of Y are dependent upon the values of X. The X variable may or may not be causing change in the Y variable. For example, while estimating sales of a product from figures on However, there may or may not be causal connection between these two sales. In fact, in certain cases, the cause-relation may be just opposite of what appears to be the obvious one.

USES OF REGRESSION ANALYSIS

In economics it is the basic technique for measuring or estimating the relationship among economic variables that constitute the essence of economic theory and economic life. For example, if we know that two variables, price (X) and demand (Y), are closely related we can find out the most probable value of X for a given value of Y or the most probable value of Y for a given value of X. Similarly, if we know that the amount of tax and the rise in the price of a commodity are closely related, we find the study of regression is of considerable help to the economists and businessmen. The uses of regression are not confined to economics and business field only. Its applications are extended to almost all the natural, physical and social sciences. The regression analysis attempts to accomplish the following:

1. Regression analysis provides estimates of values of the dependent variable from values of the independent variable. The device used to accomplish this estimation procedure is the *regression line.* The regression line describes the average relationship existing between x and Y variables, *i.e.*, it displays mean values of X for given values of Y. The equation of this line, known as the regression equation, provides estimates of the dependent variable when values of the independent variable are inserted into the equation.

2. With the help of regression coefficients we can calculate the correlation coefficient. The square of correlation coefficient (r), called coefficient of determination, measures the degree of association of correlation that exists between the two variables. It assesses the proportion of variance in the dependent variable that has been accounted for by the

regression equation. In general, the greater the value of r^2 the better is the fit and the more useful the regression equations as a predictive device.

3. A second goal of regression analysis is to obtain a measure of the error involved in using the regression line as a basis for estimation. For this purpose the standard error of estimate is calculated. This is a measure of the scatter or spread of the observed values of Y around the corresponding values estimated from the regression line. If the line fits the data closely, that is, if there is little scatter of the observations around the regression line, good estimates can be made of the Y variable. On the other hand, if there is a great deal of scatter of the observations around the fitted regression line, the line will not produce accurate estimates of the dependent variable.

DIFFERENCE BETWEEN CORRELATION AND REGRESSION ANALYSIS

Correlation and regression analysis are constructed under different assumption they furnish different types of information and it is not always clear as to which measure should be used in a given problem situation. The following are the points of difference between correlation and regression:

1. In correlation analysis r_{xy} is a measure of direction and degree of linear relationship between two variables X and Y, r_{xy} and r_{yx} are symmetric ($r_{xy} = r_{yx}$), *i.e.*, it is immaterial which of X and Y is dependent variable and which is independent variable. In regression analysis the regression coefficients b_{xy} and b_{yx} are not symmetric, *i.e.*, $b_{xy} \neq b_{yx}$ and hence it definitely makes a difference as to which variable is dependent and which is independent.

2. There may be nonsense correlation between two variables which is purely due to chance and has no practical relevance such as increase in income and in crave in weight of a group of people. However, there is nothing like non sense regression

3. Correlation is merely a tool of ascertaining the degree of relationship between two variables and, therefore, we cannot say that one variable is the cause and other the effect. For example, a high degree of correlation between price and demand for a certain commodity or a particular point of time may not suggest which is the cause and which is the effect. However, in regression thus making it possible to study the cause and effect relationship. It should be noted that the presence of association does not imply causation, but the existence of causation always implies association. Statistical evidence can only establish the

presence or absence of association between variables whether causation exists or not depends purely on reasoning.

4. Whereas coefficient is a measure of degree of covariability between X and Y, the objective of regression analysis is to study the 'nature of relationship between the variables so that we may be able to predict the value of one on the basis of another. The closer the relationship between two variables, the greater the confidence that may be placed in the estimates.

REGRESSION LINES

The regression line of Y on X gives the most probable values of Y for given values of X and the regression line of X on Y gives the most probable values of X for given values of Y. However, when there is either perfect positive or perfect negative correlation between the two variables ($r = \pm 1$) the regression lines will coincide, *i.e.*, we will have only one line. The farther the two regression lines from each other. The lesser is the degree of correlation and the nearer the two regression lines to each other, the higher is the degree of correlation. If the variables are independent, r is zero and the lines of regression are at right angles, *i.e.*, parallel to OX and OY.

It should be noted that the regression lines cut each other at the point of average of X and Y, *i.e.*, if from the point where both the regression lines cut each other a perpendicular is drawn on the X-axis, we will get the mean value of X and if from that point a horizontal line is drawn on the Y-axis, we will get the mean value of Y.

It is important to note that the regression lines are drawn on least squares assumption which stipulates that the sum of squares of the deviations of the observed 'Y' values from the fitted line shall be minimum. The total of the squares of the deviations of the various points is minimum only form the line of best fit. The deviation from the points from the line of best fit can be measured in two ways–vertical, *i.e.*, parallel to Y-axis, and horizontal, *i.e.*, parallel to X-axis. For minimising the total of the squares separately it is essential to have two regression lines. The regression line of Y on X is drawn in such a way that it minimises total of squares of the vertical deviations and the regression line of X on Y minimises the total squares of the horizontal deviations example:

Height of fathers(inches)	65	63	67	64	68	62	70	66	68	67	69	71
Height of sons (inches)	68	66	69	65	69	66	68	65	71	67	68	70

The two regression equations corresponding to these variables are:

$$X = -3.38 + 1.036\ Y \qquad ...(i)$$

$$y = 35.82 + 2.476\ X \qquad ...(ii)$$

By assuming any values of Y we can find out corresponding values of X from Eq. (i).

For example if Y = 65,

X would be –3.38 + 1.036(65) = 63.96

Similarly, if Y = 70,

X would be – 3.38 + 1.036(70) = 69.14

We can plot these points on the graph and obtain regression line of X on Y:

Similarly, by assigning any values to X is Eq. (ii) we can obtain corresponding values of Y. Thus, if X 63, Y would be:

35.82 + .476(63) = 65.808 or 65.81 and for X = 70,

Y would be 35.82 + .476(70) = 69.14

The graph of original data and these lines would be as follows:

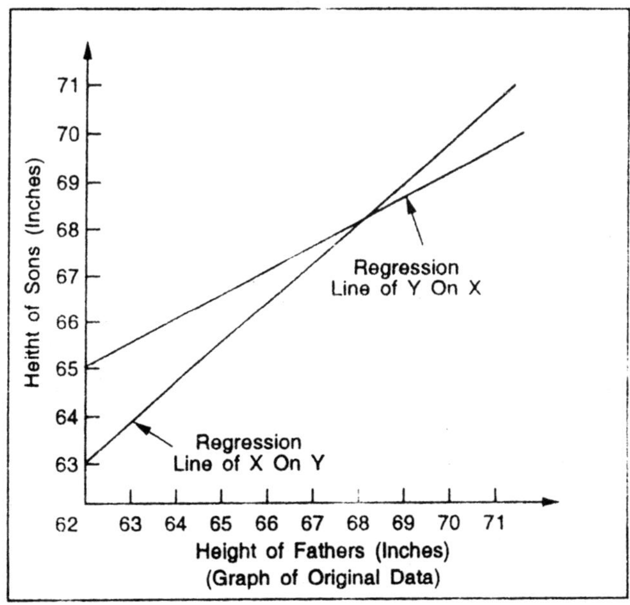

(Graph of Original Data)

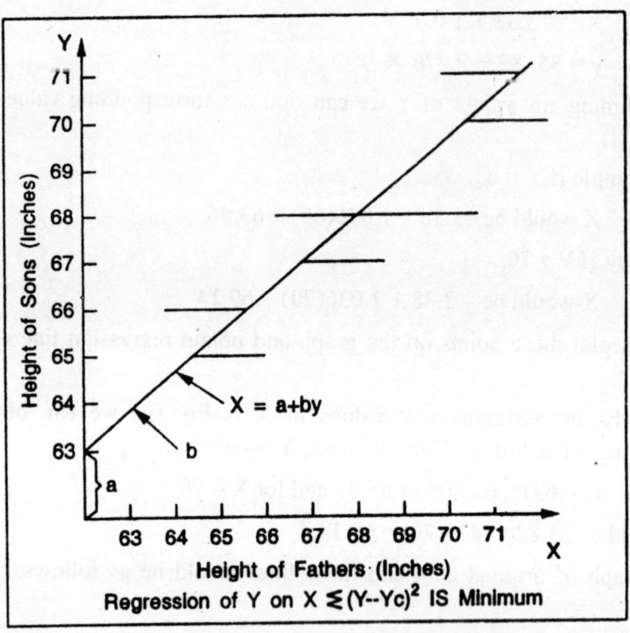

• Regression of Y on X, $\sum (Y - Y_c)^2$ is minimum,

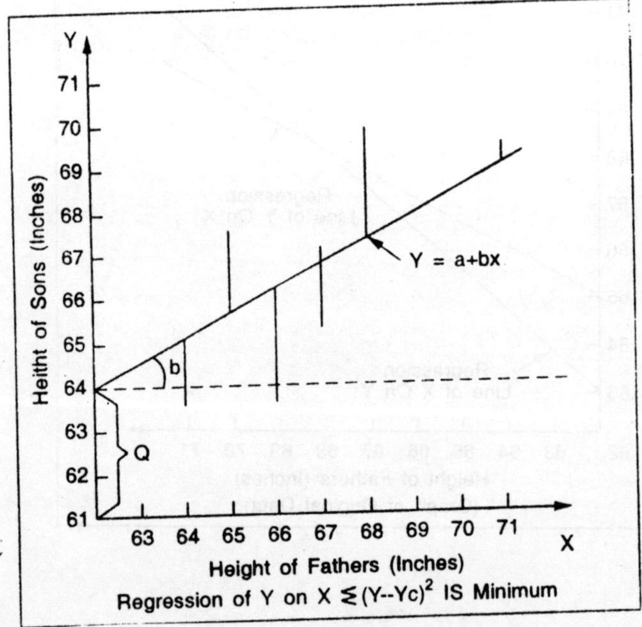

REGRESSION EQUATIONS

Regression equation, also known as estimating equations, are algebraic expression of the regression lines. Since there are two regression lines, there are two regression equations–the regression equation of X on Y is used to describe the variations in the values of X for given changes in Y and the regression equation of Y on X is used to describe the variation in the values of Y for given changes in X.

Regression Equation of Y on X

The regression equation of Y on X is expressed as follows:

$$Y = a + bX$$

It may be noted that in this equation 'Y' is a dependent variable, *i.e.*, its value depends on X, 'X' is independent variable, *i.e.*, we can take a given value of X and compute the value of Y.

'a' is "Y intercept" because its value is the point at which the regression line crosses the Y-axis, that is, the vertical axis. 'b' is the "slope" of line. It represents change in Y variable for a unit change in X variable.

'a' and 'b' in the equation are called numberical constants because for any given straight line, their value does not change.

If the values of the constants 'a' and 'b' are obtained, the line is completely determined. But the question is how to obtain these values. The answer is provided by the method of Least Squares which states that the line should be drawn thought the plotted points in such a manner that the sum of the squares of the deviations of the actual Y values from the computed Y values is the least, or in other words, in order to obtain a line which fits the points best $\Sigma(Y - Y_c)^2$, should be minimum. Such a line is known as the line of 'best fit'.

A straight line fitted by least squares has the following characteristics:

1. The straight line goes though the overall mean of the data $(\overline{X}. \overline{Y})$.
2. The deviations above the line equal those below the line, on the average. This means that the total of the positive and negative deviations is zero, or $\Sigma(Y - Y_c) = 0$.
3. It gives the best fit to the data in the sense that it makes the sum of the squared deviations from the line, $\Sigma(Y - Y_c)^2$, smaller than they would be from any other straight line. This property accounts for the name 'Least Squares'.
4. When the data represent a sample from a large population the least squares line is a 'best' estimate of the population regression line with

a little algebra and differential calculus it can be shown that the following two equations. If solved simultaneously, will yield values of the parameters a and b such that the least squares requirement is fulfilled:

$$\Sigma Y - Na + b\Sigma X$$

$$\Sigma XY = a\Sigma X + b\Sigma X^2$$

These equitations are usually called the *normal equations.* In the equations ΣX, ΣXY, ΣX^2 indicate totals which are computed from the observed pairs of values of two variables X and Y to which the least squares estimating line is to be fitted and N is the number of observed pairs of values.

Regression Equation of X on Y

The regression equation of X on Y is expressed as follows:

$$X_c = a + bY$$

To determine the values of a and b the following two normal equations are to be solved simultaneously:

$$\Sigma X = Na + b\Sigma Y$$

$$\Sigma XY = a\Sigma Y + b\Sigma Y^2$$

Example 1:

From the following data obtain the two regression equations:

X	*6*	*2*	*10*	*4*	*8*
Y	*9*	*11*	*5*	*8*	*7*

Solution:

Obtaining Regression Equations

X	Y	Xy	X²	Y²
6	9	54	36	81
2	11	22	4	121
10	5	50	100	25
4	8	32	16	64
8	7	56	64	49
$\Sigma X = 30$	$\Sigma Y = 40$	$\Sigma XY = 214$	$\Sigma X^2 = 220$	$\Sigma Y^2 = 340$

Regression equation of Y on X: $Y_c = a + bX$

To determine the values of a and b we have to solve the following equations.

$$\Sigma Y = Na + b\Sigma X$$

$$\Sigma XY = a\Sigma X + b\Sigma X^2$$

Substituting the values $40 = 5a + 30b$ ...(i)

$$214 = 30a + 220b \quad ...(ii)$$

Multiplying equation (i) by 6. $240 = 30a + 180b$ ...(iii)

$$214 = 30a + 220b \quad ...(iv)$$

Deducting equation (iv) from (iii) $-40b = 26$ or $b - 0.65$

Substituting the value of b in equation (i).

$$40 = 5as + 30(-0.65) \text{ or } 5a = 40 + 19.5 = 59.5 \text{ or } a = 11.9$$

Putting the values of a and b in the equation, the regression of Y on X is

$$Y = 11.9 - 0.65 X$$

Regression line of X on Y : $Xc = a + bY$

and the two normal equations are:

$$\Sigma X = Na + b\Sigma Y$$

$$\Sigma XY = a\Sigma Y + b\Sigma Y^2$$

$$30 = 5a + 40b \quad ...(i)$$

$$214 = 40a + 340b \quad ...(ii)$$

Multiplying equation (i) by 8; $240 = 40a + 320b$ ...(iii)

$$214 = 40a + 340b \quad ...(iv)$$

From Eqn. (iii) and (iv)$-20b = 26$ or $b = -1.3$

Substituting the value of b in equation (i); $30 = 5a + 40(-1.3)$

$$5a = 30 + 52 = 82 \quad \therefore a = 16.4$$

Putting the values of a and b in the equation, the regression line of X on Y is $X = 16.4 - 1.3 Y$.

Deviations Taken from Arithmetic Means of X and Y

The above method of finding out regression equation is tedious. The calculations can very much be simplified if instead of dealing with the actual values of X and Y we take the deviations of X and Y series from their respectively means. In such a case the two regression equations are written as follows:

(i) Regression Equation of X on Y; $X - \overline{X} = r\frac{\sigma_x}{\sigma_y}(Y - \overline{Y})$

$\overline{X}$ is the mean of X series; $\overline{Y}$ is the mean of Y series

$r\frac{\sigma_x}{\sigma_y}$ is known as the regression coefficient of X on Y,

The regression coefficient of X on Y is denoted by the symbol b_{xy} or b_1. It measures the change in X corresponding to a unit change in Y, When deviations are taken from the means of X and Y, the regression coefficient of X on Y is obtained as follows:

$$b_{xy} \text{ or } r\frac{\sigma_{x*}}{\sigma_y} = \frac{\Sigma xy}{\Sigma y^2}$$

Instead of finding out the value of correlation coefficient, σ_x, σ_y, ete, we can find the value of regression coefficient by calculating Σxy and Σy^2 and dividing the former by the latter.

(ii) Regression Equation of Y on X; $Y - \overline{Y}\ r\frac{\sigma_y}{\sigma_x}(X - \overline{X})$

$r\frac{\sigma_y}{\sigma_x}$ is the regression coefficient of Y on X. It is denoted by b_{yx} or b_2. It measures the change in Y corresponding to a unit change in X. When deviations are taken from actual means, the regression coefficient of Y on X can be obtained as follows:

$$r\frac{\sigma_{y*}}{\sigma_x}\ \frac{\Sigma xy}{\Sigma x^2}$$

It should be noted that the under-root of the product of two regression coefficients gives us the value of correlation coefficient. Symbolically,

$$r = \sqrt{b_{xy} \times b_{yx}}$$

Proof:

$$b_{xy} = r\frac{\sigma_x}{\sigma_y} \text{ and } b_{yx} = r\frac{\sigma_y}{\sigma_x}$$

$$b_{xy} \times b_{yx} = r\frac{\sigma_x}{\sigma_y} \times r\frac{\sigma_y}{\sigma_x} = r^2 \qquad \therefore\ r = \sqrt{b_{xy} \times b_{yx}}$$

The following points should be noted about the regression coefficients:

1. Both the regression coefficients will have the same sign., *i.e.*, either they will be positive or negative. It is never possible that one of the regression coefficients is negative and the other positive.

2. Since the value of the coefficients of correlation (r) cannot exceed one, one of the regression coefficients must be less than one or. in other words, both the regression coefficients cannot be greater than one. For example, if $b_{xy} = 1.2$ and $b_{yx} = 1.4$ the value of correlation coefficient would be $\sqrt{1.2 \times 1.4} = 1.296$ which is not possible.

3. The coefficients of correlation will have the same sign as that of regression coefficients, *i.e.*, if regression coefficients have a negative sign, r will also be negative; and if regression coefficients have a positive sign, r would also be positive. For example, if $b_{xy} = -0.8$ and $b_{xy} = -1.2$ r would be $\sqrt{-0.8 \times -1.2} = -098$ and not -0.98.

4. Since $b_{xy} = r\frac{\sigma_x}{\sigma_y}$, we can find out any of the four values given the other three. For example, if we know that $r = 0.6$, $\sigma_x = 4$ and $b_{xy} = 0.8$, we can find σ_y.

$$b_{xy} = r\frac{\sigma_x}{\sigma_y}$$

Substituting the given values: $0.8 = \frac{0.6 \times 4}{\sigma_y}$ or $\sigma_y = \frac{2.4}{0.8} = 3.$

5. Regression coefficients are independent of change of origin but not of scale.

Proof:

Let $U = X - A$ and $V = Y - B$

wher A and B are arbitrary means

$$X = A + U \qquad\qquad Y = B + V$$

$$\overline{X} = a + \overline{U} \qquad\qquad \overline{Y} = B + \overline{V}$$

$$(X - \overline{X}) = (U - \overline{U}) \qquad\qquad (Y - \overline{Y}) = (V - \overline{V})$$

$$\sigma_x^2 = \sigma_u^2 \qquad\qquad \sigma_y^2 = \sigma_\upsilon^2$$

$$\Sigma(x - \overline{x})(y - \overline{y}) = \Sigma(u - \overline{u})(\upsilon - \overline{\upsilon})$$

Hence change of origin has no effect on the regression coefficient.

Example 2:

For certain X and Y series which are correlated, the two lines of regression are:

$$5X - 6Y + 90 = 0$$

$$15X - 8Y - 130 = 0$$

Find the means of the two series and the correlation coefficient.

Solution:

(i) Finding mean of the two series:

$$5X - 6Y = -90$$

$$15X - 8Y = 130$$

Multiplying eq. (i) by 3, $15X - 18Y = -270$

$$15X - 8Y = 130$$

$$- \quad + \quad -$$

$$-10Y = -400$$

$$Y = 40 \text{ or } \overline{Y} = 40$$

Putting the value of Y in eq. (i) $5X - 6(40) = -90$

$$5X = -90 + 240$$

$$5X = -150 \text{ or } X = 30 \text{ or } \overline{X} = 30$$

(ii) Finding correlation coefficient. Let us assume that eq. (i) is the regression equation of X on Y;

$$5X = 6Y - 90$$

$$X = \frac{6}{5} Y - 18 \text{ or } b_{xy} = \frac{6}{5}$$

Taking eq. (ii) as the eq. or Y on X; $-8y = -15X + 130$

$$8Y = 15X - 130$$

$$Y = \frac{15}{8} X - \frac{130}{8} \text{ or } b_{yx} = \frac{15}{8}$$

Since both the regression coefficients are exceeding one, our assumption is wrong.

Hence equ. (i) is the regression eq. of Y on X

$$\therefore \quad -6Y = -5X - 90 \quad \text{or } 6Y = 5X + 90$$

or $$Y = \frac{5}{6} X + 15 \quad \text{or} \quad b_{yx} = \frac{5}{6}$$

E. (ii) is the regression eq. of X on Y; $15X = 130 + 8Y$

$$X = \frac{130}{15} + \frac{8}{15} Y; \; b_{yx} = \frac{8}{15}$$

$$r = \sqrt{b_{xy} \times b_{yx}} = \sqrt{\frac{8}{15} \times \frac{5}{6}} = 0.667$$

Example 3:

The following data relate to the scores obtained by 9 salesmen of a company in an intelligence lest and their weekly sales in thousand rupees.

Salesmen Intelligence:	*A*	*B*	*C*	*D*	*E*	*F*	*G*	*H*	*I*
Test Scores:	50	60	50	60	80	50	80	40	70
Weekly Sales:	30	60	40	50	60	30	70	50	60

(a) Obtain the regression equation of sales on intelligence test scores of the sales.

(b) If the intelligence test score of a salesman in 65, what would be his expected weekly sales?

Solution:

Let intelligence test score be denoted by X and weekly sales by Y.

Calculation of Regression Equations

X	(X – 60) x	x^2	Y	(Y – 50) y	y^2	xy
50	–10	100	30	–20	460	+200
60	0	0	60	+10	100	0
50	–10	100	40	–10	100	+100
60	0	0	50	0	0	0
80	+20	400	60	+10	100	+200
50	–10	100	30	–20	400	+200
80	+20	400	70	+20	400	+400
40	–20	400	50	0	0	0
70	+10	100	60	+10	100	+100
$\Sigma X = 540$	$\Sigma x = 0$	$\Sigma x^2 = 1600$	$\Sigma Y = 450$	$\Sigma y = 0$	$\Sigma y^2 = 1600$	$\Sigma xy = 1200$

Fregession equation of on X; $Y - \overline{Y} = r\frac{\sigma_y}{\sigma_x}(X - \overline{X})$

$$r\frac{\sigma_y}{\sigma_x} = \frac{\Sigma xy}{\Sigma x^2} = \frac{1200}{1600} = 0.75$$

$$\overline{X} = \frac{\Sigma X}{N} = \frac{540}{9} = 60, \overline{Y} = \frac{\Sigma Y}{N} = \frac{450}{9} = 50$$

$Y - 50 = 0.75 (X - 60)$

$Y - 50 = 0.75X - 45$ or $Y = 5 + 0.75X$

Expected weekly sales when intelligence test score of a salesman is 65

$$Y = 5 + 0.75X. \quad \text{Putting } X = 65$$

$$Y = 0.75 \times (65) + 5 = 48.75 + 5 = 53.75$$

Example 4:

In a partially destroyed laboratory record of an analysis of correlation data, the following results only are legible:

Variance of X = 9

Regression equations 8 X – 10Y + 66 = 0

40 X – 18 Y = 214

Find on the basis of the above information

(i) The mean values of X and Y

(ii) Coefficient of correlation between X and Y, and

(iii) Standard deviation of Y.

Solution:

(i) The mean values of X and Y $8X - 10Y = -66$ ...(i)

$40X - 18Y = 214$ ...(ii)

Multiplying equation (i) by 5 $\quad 40X - 50Y = -330$

$$40X - 18Y = 214$$

$$- \quad + \quad -$$

$$-32Y = -544$$

$$y = 17 \text{ or } \overline{Y} = 17$$

Substituting the value of Y in eq. (i); $8X - 10 \times 17 = -66$

$$8X = -66 + 170$$

$$8X = 104 \quad \therefore \quad X = 13 \text{ or } X = 13$$

(ii) For finding out the correlation coefficient, we will have to find out the regression coefficient. Since we do not know which of the two regression equations is the equation of X on Y, we make an assumption. Let us take eq. (i) as the regression equation of X on Y

$$8X = -66 \times 10Y$$

$$X = -\frac{66}{8} + \frac{10}{18}Y; \quad \text{or } b_{xy} = \frac{10}{8} = 125$$

From eq. (ii) we can calculate b_{yx} $\quad 40X - 18Y = 214$

or, $-18\ Y = 214 - 40\ X$

$$Y = -\frac{214}{18} + \frac{40}{18} X \quad \text{or} \quad b_{yx} = \frac{40}{18}$$

Since both the regression coefficients are exceeding 1, our assumption is wrong. Hence the first equation is equation of Y on X.

From eq. (i) $\qquad -Y = -8X - 66$

$$Y = -\frac{8}{10} X + 6.6 \quad \text{or} \quad b_{yx} = \frac{8}{10}$$

From eq. (ii) $\qquad b_{xy} = \frac{18}{40}$

$$r = \sqrt{\frac{8}{10} \times \frac{18}{40}} = \sqrt{0.36} = 0.6$$

$$\sigma_x = \sqrt{9} = 3;\ b_{xy} = r\frac{\sigma_x}{\sigma_y}$$

$$45 = 6\frac{3}{\sigma_y} \quad \text{or } 45\ \sigma_y = 1.8 \quad \text{or } \sigma_y = \frac{1.8}{45} = 4$$

Hence standard deviation of Y is 4.

Example 5:

From the data of illustration 1, calculate the regression equations taking deviation of items from the mean of X and Y series.

Solution:

Calculation of Regression Equations

X	$(X - \overline{X})$ X	x^2	Y	$(Y - \overline{Y})$ y	y^2	xy
6	0	0	9	+1	1	0
2	–4	16	11	+3	9	–12
10	+4	16	5	–3	9	–12
4	–2	4	8	0	0	0
8	+2	4	7	–1	1	–2
$\Sigma X = 30$	$\Sigma x = 0$	$\Sigma x^2 = 40$	$\Sigma y = 40$	$\Sigma y = 0$	$\Sigma y^2 = 20$	$\Sigma xy = -26$

Regression Equation of X on Y : $X - \overline{X} = r\frac{\sigma_x}{\sigma_y}(Y - \overline{Y})$

$$r\frac{\sigma_x}{\sigma_y} = \frac{\Sigma xy}{\Sigma y^2} = \frac{-26}{20} = -1.3$$

$$\overline{X} = \frac{30}{5} = 6;\ \overline{Y} = \frac{40}{5} = 8$$

Hence $X - 6 = -1.3(Y - 8) = -1.3\ Y + 10.4$

$$X = -1.3\ Y + 16.4 \quad \text{or} \quad X = 16.4 - 1.3Y$$

Regression Equation of Y on X : $Y\overline{Y} = r\frac{\sigma_x}{\sigma_y}(X - \overline{X})$

$$r\frac{\sigma_y}{\sigma_x} = \frac{\Sigma xy}{\Sigma y^2} = \frac{-26}{40} = -0.65$$

$$y - 8 = -0.65(X - 6) = -0.65\ X + 3.9$$

$$Y = -0.65\ X + 11.9 \quad \text{or } Y = 11.9 - 0.65\ X$$

Thus we find that the answer is the same as obtained earned earlier, however, the calculation are very much simplified without the use of normal equations.

Deviations Taken from Assumed Means

When actual means of X and Y variables are in fractions the calculations can be simplified by taking the deviations from the assumed means. When deviation are taken from assumed means the entire procedure of finding regression equations remains the same–the only difference is that instead of taking deviations from actual means, we take the deviations from assumed means. The two regression equations are:

$$X - \overline{X}\ r\frac{\sigma_x}{\sigma_y}(Y - \overline{Y})$$

The value of $r = \frac{\sigma_x}{\sigma_y}$ will now be obtained as follows:

$$r\frac{\sigma_x}{\sigma_y} = \frac{N\Sigma d_x d_y - \Sigma d_x \times \Sigma d_y}{N\Sigma d_y^2 - (\Sigma d_y)^2}$$

$$d_x = (X - A) \text{ and } d_y = (Y - A)$$

Similarly, the regression equation of Y on X is

$$Y - \overline{Y} = r\frac{\sigma_y}{\sigma_x}(X - \overline{X})$$

$$r\frac{\sigma_y}{\sigma_x} = \frac{N\Sigma d_x d_y - \Sigma d_x \Sigma d_y}{N\Sigma d_x^2 - (\Sigma d_x)^2}$$

It should be noted that in both the cases the numberator is the same, the only difference is in the denominator. When the regression coefficients are calculated from correlation table their vales are obtained as follows:

$$*r\sigma_x = \frac{N\Sigma fd_x d_y - \Sigma fd_x \Sigma fd_y}{N\Sigma fd_y^2 - (\Sigma fd_y)^2} \times \frac{i_x}{i_y}$$

i_x = class interval of X variable; and

i_y = class interval of Y variable

Similarly, $$r\frac{\sigma_y}{\sigma_x} = \frac{N\Sigma fd_x d_y - \Sigma d_x \Sigma fd_y}{N\Sigma fd_x^2 - (\Sigma fd_x)^2} \times \frac{i_y}{i_x}$$

As is clear from above the formulate for calculating regression coefficients in a correlation table are the same--the only difference is that in a correlation table we are given frequencies also and hence we have multiplied every value by f.

Example 6:

From the data of illustration 1 (a) calculate regression equations by taking deviations of X series from 5 and of Y series from 7.

X	*(X –5)*		*Y*	*(Y – 7)*		
	d_x	d_x^2		d_y	d_y^2	$d_x d_y$
6	+1	1	9	+2	4	+2
2	–3	9	11	+4	16	–12
10	+5	25	5	–2	4	–10
4	–1	1	8	+1	1	–1
8	+3	9	7	0	0	0
$\Sigma X = 30$	$\Sigma X_x = +5$	$\Sigma d_x^2 = 45$	$\Sigma Y = 40$	$\Sigma d_y = 5$	$\Sigma d_y^2 = 25$	$\Sigma d_x d_y = -21$

Regression equation of X on Y : $X - \overline{X} = r\frac{\sigma_x}{\sigma_y}(Y - \overline{Y})$

$$\overline{X} = \frac{\Sigma X}{N} = \frac{30}{5} = 6; \ \overline{Y} = \frac{\Sigma Y}{N} = \frac{40}{5} = 8$$

$$r\frac{\sigma_x}{\sigma_y} = \frac{N\Sigma d_x d_y - \Sigma d_x \Sigma d_y}{N\Sigma d_y^2 - (\Sigma d_y)^2}$$

$$= \frac{5(-21) - (5)(5)}{(5)(25) - (5)^2} = -\frac{105 - 25}{125 - 25} = -\frac{130}{100} = -1.3$$

$$X - 6 = -1.3 (Y - 8)$$

$$X - 6 = -1.3 Y + 10.4 \text{ or } X = 16.4 - 1.3 Y$$

Regression Equation of Y on X : $Y - \overline{Y} = r\frac{\sigma_y}{\sigma_x}(X - \overline{X})$

$$r\frac{\sigma_y}{\sigma_x} = \frac{N\Sigma d_x d_y - \Sigma d_x \Sigma d_y}{N\Sigma d_x^2 - (\Sigma d_x)^2}$$

$$= \frac{5(-21) - (5)(5)}{(5)(45) - (5)^2} = \frac{-105 - 25}{200} = -0.65$$

$$Y - 8 = -0.65 (X - 6)$$

$$Y - 8 = -0.65 X + 3.9 \quad \text{or } Y = 11.9 - 065 X$$

It is clear from this example that answer would come out to be the same whether we take deviations from actual means or assumed means.

Greaphin Regression Lines : It is quire easy to graph the regression lines once they have been computed. All one has to do is to:

(a) choose any two values (preferably well apart) for the unknown variable on the right-hand side of the equation,

(b) compute the other variable,

(c) plot the two pairs of values, and

(d) draw a straight line through the plotted points.

Example 7:

In a correlation study the following values are obtained:

	X	*Y*
Mean	*65*	*67*
Standard Deviation	*2.5*	*3.5*
Coefficient of Correlation	–	*0.8*

Find the two regression equations that are associated with the above values.

Solution:

The two regression equations are:

Regression Equation of X on Y : $X - \overline{X} = r\frac{\sigma_x}{\sigma_y}(Y - \overline{Y})$

$\overline{X} = 65$, r = 08. $\sigma_x = 2.5$. $\sigma_y = 3.5$, $\overline{Y} = 67$

Substituting the values : $X - 65 = 8 \dfrac{2.5}{3.5} (Y - 67)$

$X - 65 = .5714 (Y - 67)$

$X - 65 = .5714 Y - 38.28$ or $X = 26.72 + 0.5714 Y$

Regression Equation of Y on X: $Y - \overline{Y} = r \dfrac{\sigma_x}{\sigma_y} (X - \overline{X})$

$Y - 67 = 8 \dfrac{3.5}{2.5} (X - 65)$

$Y - 67 = 1.12 (X - 65)$

$Y - 67 = 1.12 X - 72.8$ or $Y = - 5.8 + 1.12 X$.

Example 8:

Show graphically the regression equations of illustration 3.

Solution:

(a) Regression Line of Y on X[Y – 11.9 – 0.63 X].

(i) Let X = 2, Y = 11.9 – 0.65 (2) = 119 – 1.3 = 106.

(ii) Let X = 10, Y = 11.9 – 0.65 × 10 = 5.4.

These points and the regression line through them are shown on the graph on the next page.

(b) Regression line of X on Y (X = 16.4 – 1.3 Y)

(i) Let Y = 10

∴ x = 16.4 – 1.3 (10) = 16.4 – 13 = 3.4

(ii) Let Y = 6

∴ x = 16.4 – 1.3 (6) = 16.4 – 7.8 = 8.6

Example 9:

For 50 students of a class the regression equation of marks in Statistics (X) on the marks in Accountancy (Y) is 3Y – 5X + 180 = 0. The mean marks in Accountancy is 44 and variance of marks in Statistics is 9/16th of the variance of marks in Accountancy. Find the mean marks in Statistics and the coefficient of correlation between marks in the two subjects.

Solution:

We are given

$3Y - 5X + 180 = 0$ or $3Y + 180 = 5X$

X represents marks in Statistics and Y marks in Attractancy. When Y = 44, X will be given by

$$5X = (3)(44) + 180 = 0;\ 5X = 132 + 180 \text{ or } X = \frac{312}{5} = 62.4$$

Hence the mean marks in Statistics are 62.4

For calculating coefficient of correlation we know that

$$b_{xy} = r\frac{\sigma_x}{\sigma_y}$$

Regression coefficient of X on Y from the given equation is

$$5X = 3Y + 180 \text{ or } X = 0.6\,Y + 36$$

$$\therefore \quad b_{xy} = 0.6;\ r\frac{\sigma_x}{\sigma_y} = \frac{\sqrt{9}}{\sqrt{16}} \text{ given}$$

$$\therefore \quad 0.6 = r\frac{\sqrt{9}}{\sqrt{16}} \quad \text{or} \quad 0.6 = r\frac{3}{4}$$

Hence $3r = 24 \quad \therefore \quad r = +0.8$

Example 10:

The following table shows the ages (X) and blood pressure (Y) of 8 persons.

X:	*52*	*63*	*45*	*36*	*72*	*65*	*47*	*25*
Y:	*62*	*53*	*51*	*25*	*79*	*43*	*60*	*33*

Obtain the regression equation of Y on X and find the expected blood pressure of a person who is 49 years old.

Solution:

Calculating Regression Equation of Y on X

X	(X – 50) d_x	d_x^2	Y	(Y – 50) d_y	d_y^2	d_xd_y
52	+2	4	62	+12	144	+24
63	+13	169	53	+3	9	+39
45	–5	25	51	+1	1	–5
36	–14	196	25	–25	625	+350
72	+22	484	79	+29	841	+638
65	+15	225	43	–7	49	–105
47	–3	9	60	+10	100	–30
25	–25	625	33	–17	289	+425
$\Sigma x = 405$	$\Sigma d_x = 5$	$\Sigma d_x^2 = 1737$	$\Sigma Y = 406$	$\Sigma d_y^2 = 6$	$\Sigma d_t = 2058$	$\Sigma d_xd_y = 1336$

$$Y - \overline{Y} = r\frac{\sigma_y}{\sigma_x}(X - \overline{X})$$

$$\overline{Y} = \frac{\Sigma Y}{N} = \frac{406}{8} = 50.75;\ \overline{X} = \frac{\Sigma X}{N} = \frac{405}{8} = 50.625$$

$$r\frac{\sigma_y}{\sigma_x} = \frac{N\Sigma d_x d_y - \Sigma d_x \Sigma d_y}{N\Sigma d_x^2 - (\Sigma d_x)^2} = \frac{(8)(1363) - (5)(6)}{(8)(1737) - (5)^2} = \frac{10688 - 30}{13896 - 25} = 0.768$$

$$Y - 50.75 = 0.768\,(X - 50.625)$$

$$Y - 50.75 = 0.768\,X - 38.88 \text{ or } Y = 11.87 + 0.768\,X$$

$$Y_{49} = 11.87 + 0.768\,(49) = 49.502$$

Thus the expected blood pressure of a person who is 49 years only shall be 49.5.

Example 11:

You are given the following data:

	X	*Y*
Arithmetic mean	*36*	*85*
Standard Deviation	*11*	*8*

Correlation coefficient between X and Y = 0.66

(i) Find the two Regression Equations

(ii) Estimate the value of X when Y = 75.

(B.A. Hons. Econ. Delhi Univ. B. Com Guwahati Univ. 1998)

Solution:

(i) Regression Equation of X on Y : $X - \overline{X} = r\frac{\sigma_x}{\sigma_y}(Y - \overline{Y})$

$$\overline{X} = 36,\ r = 0.66,\ \sigma_x = 11,\ \sigma_y = 8,\ \overline{Y} = 85$$

$$X - 36 = 0.66\ \frac{11}{8}\ (Y - 85)$$

$$X - 36 = 9075\ (Y - 85)$$

$$X = 9075\ Y - 77.1375 + 36 \text{ or } X = -41.1375 + 9075Y$$

Regression Equation of Y on X : $Y - \overline{Y} = r\frac{\sigma_y}{\sigma_x}(X - \overline{X})$

$$Y - 85 = .66\ \frac{11}{8}\ (X - 36)$$

$$Y - 85 = .48\ (X - 36)$$

$$Y - 85 = .48\ X - 17.82 \quad \text{or} \quad Y = 67.72 + 0.48\ X$$

(ii) From the regression equation of X on Y, we can find out the estimated value of X when Y = 75; X = .9075 (75) – 41.1375

$$= 68.0625 - 41.1375 = 26.925 \quad \text{or} \quad Y_{75} = 26.925.$$

Regression Equations in case of Correlation Table

When we are given a correlation table and we are to find out regression equation, the procedure will remain the same as discussed above. However, for finding the regression equation of Y on X and X on Y the convenient form will be $y - \overline{Y}\ b_{yx}\ (X - \overline{X})$ and $X - \overline{X} = b_{xy}\ (Y - \overline{Y})$. It may be noted that the regression coefficients are independent of origin but not of scale and hence necessary adjustment must be made. The following example will illustrate the procedure.

Example 12:

Your are given below the following information about advertising and sales:

	Ad. Exp. (X)	*Sales (Y)*
Mean	*10*	*90*
S.D	*3*	*12*

Correlation coefficient = 0.8,

(a) Calculate the two regression lines.

(b) Find the likely sales when advertisement expenditure is Rs. 15 Crore.

(c) What should be advertisement expenditure if the company wants to attain sales target of Rs. 120 Crore?

Solution:

(a) Regression Equation of X on Y : $X - \overline{X} = r\frac{\sigma_x}{\sigma_y}(Y - \overline{Y})$

$$\overline{X} = 10;\ r = 0.8,\ \sigma_x = 3,\ \sigma_y = 12,\ \overline{Y} = 90.$$

Substituting the values, $X - 10 = 8\ \frac{8}{12}\ (Y - 90)$

$$X - 10 = 0.2\ (Y - 90)$$

$$X - 10 = 0.2\ Y - 18$$

or $X = -8 + 0.2\,Y$

Regression Equation of Y on X : $Y - \overline{Y} = r\frac{\sigma_y}{\sigma_x}(X - \overline{X})$

$$Y - 90 = 0.8\ \frac{12}{3}\ (X - 10)$$

$$y - 90 = 3.2\,(X - 10)$$

$$Y - 90 = 3.2\,X - 32$$

or $Y = 58 + 3.2\,X$

(b) For finding the likely sales when advertisement expenditure is Rs. 15 Crore, we put X = 15 in eqn. (ii).

$$Y = 3.2\,(15) + 58$$

$Y = 46 + 58 = 106$ or Thus $Y_{15} = 106$ lakhs.

(c) For determining advertisement expenditure for a sales target of Rs. 120 lakhs, put Y = 120 in eqn. (i).

$$X = 0.2\,(120) - 8$$

or $X = 24 - 8 = 16.$

Hence X_{120} = Rs. 16 corores.

Example 13:

(a) Given $b_{xy} = 85$, $b_{yx} = 89$ and the standard deviation of $X = 6$, find the value of 'r' and σ_y.

(b) The following data are given for marks in Company Law and Statistics in a certain year:

Mean Marks in Statistics	*39.5*
Mean Marks in Company Law	*47.6*
Standard deviation of Marks in Statistics	*10.8*
Standard deviation of Marks in Company Law	*16 9*
'r' between marks in Statistics and Company Law	*0.42*

Determine the two equations of regression. Also calculate expected marks in Company Law of a candidate who obtained 50 marks in Statistics.

Solution:

(a) The correlation coefficient is the under-root of the two regression coefficients *i.e.*, $r = \sqrt{b_{yx} \times b_{xy}}$; $b_{xy} = 85$; $b_{yx} = .89$.

Hence $r = \sqrt{85 \times 89} = 0.87$

$$b_{xy} = r\frac{\sigma_x}{\sigma_y}$$

$$85 = 87\frac{6}{\sigma_y}$$

$$85_{\sigma_y} = 5.22 \text{ or } \sigma_y = 6.14.$$

(b) Let the marks in Statistics be denoted by X and marks in Company Law by Y.

Regression equation of X on Y : $X - \overline{X} = r\frac{\sigma_x}{\sigma_y}(Y - \overline{Y})$

$$\overline{X} = 39.5,\ \overline{Y}\ 47.6,\ \sigma_x = 10.8,$$

$$\sigma_y = 16.9,\ r = +0.42$$

$$X - 39.5 = 0.42\frac{10.8}{16.9}(Y - 47.6)$$

$$X = 39.5 = 268\,Y - 12.757$$

or $X = 26.743 + 0.268\,Y$

Regression equation of Y on X : $Y - \overline{Y} = r\frac{\sigma_y}{\sigma_x}(X - \overline{X})$

$$Y - 47.6 = 0.42\frac{16.9}{10.8}(X - 39.5)$$

$$X - 47.6 = 0.657\,X - 25.96$$

or $Y = 21.64 + 0.657\,X$

If X is 50, Y will be = 0.657 (50) + 21.64 = 32.85 + 21.64 = 54.49

Thus the estimated marks in Company Law of a candidate who obtained 50 marks in Statistics are 54.49 or 54.

Example 14:

Obtain the regression equation of Y on X and the value of r from the following table giving the marks in Accountancy and Statistics:

X / *Y*	*5–15*	*15–25*	*25–35*	*35–45*	*Total*
0–10	*1*	*1*	–	–	*2*
10–20	*3*	*6*	*5*	*1*	*15*
20–30	*1*	*8*	*9*	*2*	*20*
30–40	–	*3*	*9*	*3*	*15*
40–50	–	–	*4*	*4*	*8*
Total	*5*	*18*	*27*	*10*	*60*

Solution:

Computation required for finding the regression equation

Y \ X		m	5–15 10	15–25 20	25–35 30	35–45 40				
		d_y \ d_x	– 1	0	+1	+2	f	f d_y	f d_y^2	f $d_x d_y$
0–10	m 5	– 2	+2 1	0	–	–	2	–4	8	+2
10–20	15	– 1	+3 3	–0 ← 6	–5 ← 5	– 2 ←1	15	–15	15	–4
20–30	25	0	0 1	0	0 8	0 ← 9	20 ← 2	0	0	0
30–40	35	1		0 ←3	+9 ←9	+6 ←3	15	+15	15	+15
40–50	45	2			+8 4	+16 4	8	+16	32	+24
		f	5	18	27	10	N = 60	Σfd_y = 12	Σfd_y^2 = 70	$\Sigma fd_x d_y$ = 37
		f d_x	– 5	0	+27	+20	Σfd_x = 42			
		f d_x^2	5	0	27	40	Σfd_x^2 = 72			
		f $d_x d_y$	+5	0	+12	+20	$\Sigma fd_x d_y$ = 37			

Regression Equation of Y on X : $Y - \overline{Y}' = r\dfrac{\sigma_y}{\sigma_x}(X - \overline{X})$

$$r\frac{\sigma_y}{\sigma_x} = \frac{N\Sigma fd_x d_y = (\Sigma fd_x)(\Sigma fd_y)}{N\Sigma fd_x^2 - (\Sigma fd_x)^2} \times \frac{i_\sqrt{y}}{i_x}$$

$$= \frac{60(37) - (42)(12)}{60(72) - (42)^2} \times \frac{10}{10} = \frac{2220 - 504}{4320 - 1764} = 0.67$$

$$\overline{Y} = A + \frac{\Sigma fd_y}{N} \times i, \text{ where } A = 20, \Sigma fd_x = 12, N = 60, i = 10$$

$$\overline{Y} + 25 + \frac{12}{60} \times 10 = 27$$

$$\overline{X} = A + \frac{\Sigma fd_x}{N} \times i, \text{ where } A = 20, \Sigma fd_x = 42, N = 60, i = 10$$

$$\overline{X} = 20 + \frac{42}{60} \times 10 = 27$$

$$Y - 27 = 0.67\,(X - 27) = 0.67\,Y - 18.09$$

or $Y = -\ 18.09$ or $Y = 8.91 + 0.67\,X$

Regression Equation of X on Y : $X - \overline{X} = r\frac{\sigma_x}{\sigma_y}(Y - \overline{Y})$

$$r\frac{\sigma_x}{\sigma_y} = \frac{N\Sigma fd_x d_y - \Sigma fd_x \Sigma fd_y}{N\Sigma fd_y^2 - (\Sigma fd_y)^2} \times \frac{i_x}{i_y}$$

$$\frac{60(37) - 42 \times 12}{60(70) - (12)^2} = \frac{2220 - 504}{4200 - 144} = 0.423$$

$$X - 27 = 0.423\,(Y - 27) = 0.423Y - 11.42$$

or $X = 15.58 + 0.423\,Y$

Correlation Coefficient : $r = \sqrt{b_{xy} \times b_{yx}} = \sqrt{0.423 \times 0.67} = 0.532$

STANDARD ERROR OF ESTIMATE

With the help of regression equations perfect prediction is practically impossible. For example, the revenue for the year from gasoline sales (Y) based on automobile registration (X) as of a certain date would no doubt be approximated fairly closely, but predication may not be exact to the nearest rupee nor probably to the nearest '000 rupees. What is needed, the, is a measure which would indicate how precise the prediction of y is, based on X or conversely, how inaccurate the predication might be. The measure is called the *standard error of estimate.* The standard error of estimate, symbolised by S_{yx} is the same concept as the standard deviation discussed in Chapter 8, The standard deviation measures the dispersion about an average, such as the mean. The standard error of estimate measures the dispersion about an average line, called the regression line. The formula for calculating the standard error of estimate is:

$$S_{yx} = \frac{\sqrt{\Sigma(Y - \overline{Y}_c)^2}}{N} \text{ i.e., } \frac{\sqrt{\text{Unexplained Variation}}}{N}$$

also $$S_{y.x} = \sigma_y \sqrt{1 - r^2}$$

where $S_{y.x}$ = the standard error of regression of Y values from Y_c.

This formula is not convenient from the computational point of view because it requires the computation of $(Y - Y_c)$. A more convenient formula is

$$S_{y.x} = \sqrt{\frac{\Sigma Y^2 - a\Sigma Y - b\Sigma XY}{N}}$$

The standard error of regression of X values from X_x is

$$S_{xy} = \sqrt{\frac{\Sigma(X - X_c)^2}{N}} \text{ also } S_{xy} = s^x \sqrt{1 - r^2}$$

Also $$S_{xy} = \frac{\sqrt{\Sigma X^2 - a\Sigma X - b\Sigma XY}}{N}$$

S_{xy} = the standard error of regression of X values from X_c.

The standard error of estimate measures the accuracy of the estimated figures. The smaller the value of standard error of estimate, the closer will be the dots to the regression line and the better the estimates based on the equation for this line. If standard error of estimate is zero, then there is no variation about the line and the correlation will be perfect. Thus with the help of standard error of estimate it is possible for us to ascertain how good and representative the regression line is a description of the average relationship between two series.

Example 15:

Given the following data:

X	*6*	*2*	*10*	*4*	*8*
Y	*9*	*11*	*5*	*8*	*7*

Find the two regression equations and calculate the standard error of the estimate (S_{yx} and S_{yx}).

Solution:

The two regression equations are:

$$Y = 11.9 - 0.65 X \text{ and } X = 16.4 - 1.3 Y$$

From the regression equation of Y on X for various values of X we can find out the corresponding values, and from the equation of X on Y we can find out X_c. These values are as follows:

X	Y	Y_c	X_c	$(Y - Y_c)^2$	$(X - X_c)^2$
6	9	8.0	4.7	1.00	1.69
2	11	10.6	2.1	0.16	0.01
10	5	5.4	9.9	0.16	0.01
4	8	9.3	6.0	1.69	4.00
8	7	6.7	7.3	0.09	0.49
$\Sigma X = 30$	$\Sigma Y = 40$	ΣY_c 40	$\Sigma X_c = 30$	$\Sigma(Y - Y_c)^2 = 3.1$	$\Sigma(X - X_c)^2 = 6.20$

$$S_{y.x} = \sqrt{\frac{\Sigma(Y - Y_c)^2}{N}} = \frac{\sqrt{3.1}}{5} = \sqrt{0.62} = 0.787$$

$$S_{x.y} = \sqrt{\frac{\Sigma(X - X_c)^2}{N}} = \frac{\sqrt{6.2}}{5} = \sqrt{1.24} = 1.114$$

Example 16:

The following calculations have been made for prices of twelve stocks (X) on the Calcutta Stock Exchange on a certain day along with the volume of sales in thousands of shares (Y). For these calculations find the regression equation of prices on stocks, on the volume of sales of shares.

$\Sigma X = 580, \quad \Sigma Y = 370, \quad \Sigma XY = 11494,$

$\Sigma X^2 = 41658, \quad \Sigma Y^2 = 17206.$

Solution:

Regression equation of X on Y

$$X - \overline{X} = r\frac{\sigma_x}{\sigma_y}(Y - \overline{Y})$$

$$\overline{X} = \frac{\Sigma X}{N} = \frac{580}{12} = 48.333; \ \overline{Y} = \frac{\Sigma Y}{N} = \frac{370}{12} = 30.833$$

$r\frac{\sigma_x}{\sigma_y}$ or $b_{xy} = \frac{\Sigma XY - N\overline{XY}}{\Sigma Y^2 - N(\overline{Y})^2}$ [when figures are given in original values]

$$= \frac{11494 - 12(48.333 \times 30.833)}{17206 - 12(30.833)^2}$$

$$= \frac{11494 - 17883.015}{17206 - 11408.09} = \frac{-6389.015}{5797.91} = -1.102$$

$$X - 48.333 = - 1.102 (Y - 30.833)$$

$$X - 48.333 = 1.102 Y - 33.427$$

or $$X = 81.76 - 1.102 Y.$$

Example 16(a):

The following data give the experience of machine operators and their performance ratings as given by the number of good parts turned out per 100 pieces:

Operator	*1*	*2*	*3*	*4*	*5*	*6*	*7*	*8*
Experience (X) →	*16*	*12*	*18*	*4*	*3*	*10*	*5*	*12*
Performance Ratings (Y)	*87*	*88*	*89*	*68*	*78*	*80*	*75*	*83*

Calculate the regression lines of performance ratings on experience and estimate the probable performance if an operator has 7 years experience.

Solution:

Calculation Regression Line of Y on X

Experience X	**(X − $\overline{X}$) $\overline{X}$ = 10 x**	**x²**	**Rerformance Ratings Y**	**(Y − $\overline{Y}$) $\overline{Y}$ = 81 y**	**y²**	**xy**
16	+6	36	87	+6	36	+36
12	+2	4	88	+7	49	+14
18	+8	64	89	+8	64	+64
4	−6	36	68	−13	169	+78
3	−7	49	78	−3	9	+21
10	0	0	80	−1	1	0
5	−5	25	75	−6	36	+30
12	+2	4	83	+2	4	+4
ΣX = 80	Σx = 0	Σx² = 218	ΣY = 648	Σy = 0	Σy² = 368	Σxy = 247

Regression Equation of Y on X : $Y - \overline{Y} = b_{xy} (X - \overline{X})$

$$b_{yx} = \frac{\Sigma xy}{\Sigma x^2} = \frac{247}{218} = 1.133; \overline{Y} = \frac{648}{8} = 81; \overline{X} = \frac{80}{8} = 10$$

$\therefore Y - 81 = 1.133 (X - 10) = 1.133; X - 11.133$

or $Y = 69.67 + 133X$

when $X = 7$, Y will be 1.133 (7) + 69.67 = 77.601.

Thus the probable performance of an operator who has 7 years' experience = 77.601 or 78 good parts out of 100.

Example 17:

From the following data obtain the two regression equations and calculate the correlation coefficient:

X:	*1*	*2*	*3*	*4*	*5*	*6*	*7*	*8*	*9*
Y:	*9*	*8*	*10*	*12*	*11*	*13*	*14*	*16*	*15*

Estimate the value of Y which should correspond on an average to X = 6.2,

Solution:

Calculation of Regression Equations and Correlation coefficient

X	**$(X - \overline{X})$** **x**	**x^2**	**Y**	**$(Y - \overline{Y})$** **y**	**y^2**	**xy**
1	–4	16	9	–3	9	+12
2	–3	9	8	–4	16	+12
3	–2	4	10	–2	4	+4
4	–1	1	12	0	0	0
5	0	0	11	–1	1	0
6	+1	1	13	+1	1	+1
7	+2	4	14	+2	4	+4
8	+3	9	16	+4	16	+12
9	+4	16	15	+3	9	+12
$\Sigma X = 45$	$\Sigma x = 0$	$\Sigma x^2 = 60$	$\Sigma Y = 108$	$\Sigma y = 0$	$\Sigma y^2 = 60$	$\Sigma xy = 57$

Regression Equation of X on Y : $X - \overline{X} = r\dfrac{\sigma_x}{\sigma_y}(Y - \overline{Y})$

$$\overline{X} = \frac{\Sigma X}{N} = \frac{45}{9} = 5;\ \overline{Y} = \frac{\Sigma Y}{N} = \frac{108}{9} = 12$$

$$r\frac{\sigma_x}{\sigma_y} = \frac{\Sigma xy}{\Sigma y^2} = \frac{57}{60} = 0.95$$

$X - 5 = .95\ (Y - 12)$

$X - 5 = .95\ Y - 11.4$ or $X = -6.4 + 0.95\ Y$

Regression Equation of Y on X: $Y - \overline{Y} = r\dfrac{\sigma_y}{\sigma_x}(X - \overline{X})$

$$r\frac{\sigma_y}{\sigma_x} = \frac{\sigma xy}{\Sigma y^2} = \frac{57}{60} = 0.95$$

$Y - 12 = .95\ (X - 5)$

$Y - 12 = .95\ X - 4.75$ or $Y = 7.25 + 0.95\ X$

$Y_{6.2} = .95\ (6.2) + 7.25 = 13.14$

Correlation Coefficient: The correlation coefficient is the under-root of the two regression coefficients, *i.e.*, $r = \sqrt{b_{xy} \times b_{yx}}$

$b_{xy} = .95;\ b_{yx} = .95$

Hence $r = \sqrt{.95 \times .95} = 0.95$

Example 18:

Obtain regression equation of Y on X and estimate Y when X = 55 from the following:

X:	*40*	*50*	*38*	*60*	*65*	*50*	*35*
Y:	*38*	*60*	*55*	*70*	*60*	*48*	*30*

Solution:

Calculation of Regression Equation of Y on X

X	**(X – 48) dx**	d_x^2	**Y**	**(Y – 50) dy**	d_y^2	d_xd_y
40	–8	64	38	–12	144	+96
50	+2	4	60	+10	100	+20
38	–10	100	55	+5	25	–50
60	+12	144	70	+20	400	+240
65	+17	289	60	+10	100	+170
50	+2	4	48	–2	4	–4
35	–13	169	30	–20	400	+260
$\Sigma X = 338$	$\Sigma d_x = 2$	$\Sigma d_x^2 = 774$	$\Sigma Y = 361$	$\Sigma d_y = 11$	$\Sigma d_y^2 = 1173$	$\Sigma d_xd_y = 732$

Regression eq. of Y on S : $Y - \overline{Y} = r\frac{\sigma_y}{\sigma_x}(X - \overline{X})$

$$\overline{Y} = \frac{\Sigma Y}{N} = \frac{361}{7} = 51.57;\ \overline{X} = \frac{\Sigma X}{N} = \frac{338}{7} = 48.29$$

$$r\frac{\sigma_y}{\sigma_x} = \frac{N\Sigma dxdy - \Sigma dxdy}{N\Sigma dx^2 - (\Sigma dx)^2}$$

$$= \frac{7(732) - (2)(11)}{7(774) - (2)^2} = \frac{5124 - 22}{5418 - 4} = 0.942$$

$$Y - 51.57 = .942\ (X - 48.29)$$

$$Y - 51.57 = 942\ X - 45.49$$

$$Y = .942\ X + 6.08$$

Estimate of Y when $X = 55$

$$Y = .942\ (55) + 6.08 = 51.81 + 6.08 = 57.89$$

Example 19:

Obtain the flies of regression from the following data:

X:	*4*	*5*	*6*	*8*	*11*
Y:	*12*	*10*	*8*	*7*	*5*

Verify that the coefficient of correlation is the geometric mean of the two coefficients of regression.

Solution:

X	(X – 6) d_x	d_x^2	Y	(Y – 8) d_y	d_y^2	d_xd_y
4	–2	4	12	+4	16	–8
5	–1	1	10	+2	4	–2
6	0	0	8	0	0	0
8	+2	4	7	–1	1	–2
11	+5	25	5	–3	9	–15
$\Sigma X = 34$	$\Sigma d_x = 4$	$\Sigma d_x^2 = 34$	$\Sigma Y = 42$	$\Sigma_{\delta\psi} = +4$	$\Sigma d_y^2 = 30$	$\Sigma d_xd_y = 27$

Regression Equation of X on Y : $X - \overline{X} = r\frac{\sigma_x}{\sigma_y}(Y - \overline{Y})$

$$\overline{X} = \frac{34}{5} = 6.8;\ \overline{Y} = \frac{42}{5} = 8.4$$

$$r\frac{\sigma_x}{\sigma_y} = \frac{N\Sigma d_x d_y - \Sigma d_x \Sigma d_y}{N\Sigma d_y^2 - (\Sigma d_y)^2}$$

$$= \frac{(5)(-27) - (4)(2)}{(5)(30) - (2)^2} = -\frac{-135 - 8}{150 - 4} = -0.979$$

$$X - 68 = -.979\ (Y - 8.4)$$

$$X - 6.8 = -.979\ Y + 8.224$$

or $$X = 15.024 - 0.979\ Y$$

Regression equation of Y on X : $Y - \overline{Y} = r\frac{\sigma_y}{\sigma_x}(X - \overline{X})$

$$Y - 8.4 = -.929\ (X - 6.8)$$

$$Y - 8.4 = -.929\ X + 6.317$$

$$Y = -.929\ X + 14.717$$

$$r = \frac{N\Sigma d_x d_y - \Sigma d_x \Sigma d_y}{\sqrt{N\Sigma d_x^2 - (\Sigma d_x)^2}\sqrt{N\Sigma d_y^2 - (\Sigma d_y)^2}}$$

$$= \frac{5(-27) - 4 \times 2}{\sqrt{5 \times 34 - (4)^2}\sqrt{5 \times 30 - (2)^2}} = \frac{-135 - 8}{\sqrt{170 - 16}\sqrt{150 - 4}}$$

$$= \frac{-143}{\sqrt{154 \times 146}} = \frac{-143}{149.947} = -0.954$$

$$r = \sqrt{b_{xy} \times b_{yx}} = -\sqrt{979 \times .929} = -0.954$$

$$r\frac{\sigma_y}{\sigma_x} = \frac{N\Sigma d_x d_y - \Sigma d_x \Sigma d_y}{N\Sigma d_x^2 - (\Sigma d_x)^2}$$

$$= \frac{(5)(-27) - (4)(2)}{5 \times 34 - (4)^2} = \frac{-135 - 8}{170 - 16} = -0.929$$

Hence the Coefficient of Correlation is in geometric mean of the two coefficients of regression.

Example 20:

Find the most likely production corresponding to a rainfall 40" from the following data:

	Rainfall	*Production*
Average	*30*	*500 kg*
Standard Deviation	*5"*	*100 kg*
Coefficient of correlation = 0.8		

Solution:

Since production depends on rainfall we denote production by Y and rainfall by X and fit a regression line of Y on X.

$$Y - \overline{Y} = r\frac{\sigma_y}{\sigma_x}(X - \overline{X})$$

$$\overline{Y} = 500,\ r = 0.8,\ \sigma_y = 100,\ \sigma_y = 5,\ \overline{X} = 30$$

Substituting the values: $Y - 500 = .8\frac{100}{5}\ (X - 30)$

$$Y = 500 = 16\ (X - 30)$$

$$Y - 500 = 16\ X - 480 \quad \text{or} \quad Y = 20 + 16\ X$$

For X = 40 Y will be: Y = 16 (40) + 20 = 660

Thus the most likely production corresponding to a rainfall of 30* is 660 kg.

Example 21:

A survey was conducted to study the relationship between expenditure on accommodation (X) and expenditure on food and entertainment (Y) and the following results were obtained:

	Mean Standard	*Deviation*
Expenditure on Accommodation	*Rs 173*	*63.15*
Expenditure on food and entertainment	*Rs. 47.8*	*22.98*
Co-efficient of Correlation	*+0.57*	

Write down the regression equation and estimate the expenditure on food and entertainment, if the expenditure on accommodation is Rs. 200.

Solution:

To solve this problem we have to fit regression equation of Y on X.

$$Y - \overline{Y} = r\frac{\sigma_y}{\sigma_x}(X - \overline{X})$$

Given $\overline{X} = 173$, $\sigma_n = 63.15$, $\overline{Y}$ 47.48, $\sigma_y = 22.98$

and r = 0.57, Substituting the value

$$Y - 47.8 = .57 \frac{22.98}{63.15} (X - 173)$$

$$Y - 47.8 = 207 (X - 173)$$

$$Y - 47.8 = .207 X - 35.81$$

$$y = 207 X + 11.99$$

Putting X = 200 in the equation

$$Y = .207 (200) + 11.99 = 53.39$$

The estimated expenditure on food and entertainment if expenditure on accommodation is Rs. 200 shall be 53.39.

Example 22:

For 10 observations on price (p) supply (S) the following data were obtained (in appropriate units):

$\Sigma p = 130$, $\Sigma S = 220$, $\Sigma p^2 = 2,288$; $\Sigma S^2 = 5,506$; $\Sigma pS = 3,467$, $N = 10$.

Obtain the line of regression of S on p and estimate the estimate the supply when the price is 16 units, and lied out the standard error of the estimate.

Solution:

Regression of S on p is given by : S = a + bp.

The value of and b can be determined by solving the following two normal equations:

$$\Sigma S = Na + \Sigma bp$$

$$\Sigma pS = a\Sigma p^2$$

Substituting the given values in the above equations

$$220 = 10a + 130b \quad ...(i)$$

$$3,467 = 130a + 2,288b \quad ..(ii)$$

Multiplying equation (i) by 13

$$2,800 = 130a + 1,690b$$

$$3,467 = 130a + 2,288b$$

$-598b = -607$ or $b = 1.015$

Substituting the value of b in equation (i)

$$220 = 10a + 130\,(1.015)$$

$$10a + 131.95 = 220 \text{ or } 10a = 220 - 131.95$$

$$\therefore \quad a = 8.805$$

Thus $\quad S = 8.805 + 1.015p$

When price is 16 units, the estimated supply will be:

$$S = 8.805 + 1.015\,(16) = 8.805 + 16.24 = 25.045$$

Standard Error of Estimate

$$S_{ps} = \sqrt{\frac{\Sigma S^2 - a\Sigma S - b\Sigma pS}{N}}$$

$$= \sqrt{\frac{5506 - 220(8.805) - 3467) - 3467(1.015)}{10}}$$

$$= \sqrt{\frac{5506 - 1937.1 - 3519}{10}} = \sqrt{\frac{49.9}{10}} = 2.234$$

Example 23:

Two random variables have the regression equation:

$$3X + 2Y - 26 = 0$$

$$6X + Y - 31 = 0$$

Find the mean values and the coefficient of correlation between X and Y. If the variable 9X = 25, find the standard deviation of Y from the data given above.

Solution:

Mean values of X and Y

$$3X + 24 = 26 \qquad \text{...(i)}$$

$$6X + Y = 31 \qquad \text{...(ii)}$$

Mulitplying eq. (i) by 2

$$6X + 4Y = 52$$

$$6X + Y = 31$$

$$- \quad - \quad -$$

$$3Y = 21$$

$$Y = 7 \quad \text{or} \quad \overline{Y} = 7$$

Putting the value of Y in eq (i)

$$3X + 14 = 26$$

or $3X = 12$ or $X = 4$

Coefficient of Correlation : For finding coefficient of correlation, we determine the value of regression coefficients from the two regression equations.

Taking eq. (i) as regression eq. of Y on X

$$2Y = 26 - 3X$$

$$Y = 13 - \frac{3}{2} X \quad \text{or } b_{yx} = \frac{-3}{2}$$

From eq. (ii) $6X = 31 - Y$

$$X = \frac{31}{6} - \frac{1}{6} Y \quad \text{or} \quad b_{xy} = \frac{-1}{6}$$

$$r = \sqrt{b_{xy} \times b_{yx}} = -\sqrt{\frac{3}{2} \times \frac{1}{2}} = -0.5$$

Standard deviation of Y

$$b_{yx} = r\frac{\sigma_y}{\sigma_x}$$

$$b_{yx} = -\frac{3}{2} \quad r = -5, \sigma_x = \sqrt{25} = 5$$

Substituting the values

$$\frac{-3}{2} = -.5\frac{\sigma_y}{5}$$

$$5X - 1.5 = -.5\sigma_y$$

or $5.\sigma_y = 7.5$ or $\sigma_y = 15$

Example 24:

The following data, based on 450 students, are given for marks in Statistics and Economics at a certain examination/

Mean marks in Statistics	*40*
Mean marks in Economics	*48*
S.D. of marks in Statistics	*12*
The variance of marks in Economics	*256*

Some of the products of deviation of marks from their respective mean 42075.

Given the equations of the two lines of regression and estimate the average marks in Economics of candidates who obtained 50 marks in Statistics.

Solution:

Let the marks in statistics be denoted by X and marks in Economics by Y.

Regression Equation of X on Y : $X \ \overline{X} = r\frac{\sigma_x}{\sigma_y}(Y - \overline{Y})$

$$\overline{X} = 40,\ \overline{Y} = 48,\ \sigma_x = 12,\ \sigma_y = \sqrt{256} = 16$$

$$r = \frac{\Sigma xy}{N\sigma_y\sigma_y} = \frac{42075}{45.0 \times 12 \times 16} = 0.487$$

$$X - 40 = .487\ \frac{12}{16}\ (Y - 48)$$

$$X - 40 = .365\ (Y - 48)$$

$$X - 40 = .365\ Y - 17.52 \quad \text{or} \quad X = 22.48 + 0.375\ Y$$

Regression Equation of Y on X : $Y - \overline{Y} = \frac{\sigma_y}{\sigma_x}(X - \overline{X})$

$$Y - 48 = .487\ \frac{16}{12}\ (X - 40)$$

$$Y - 48 = .649\ (X - 40)$$

$$Y - 48 = .649x - 25.96 \quad \text{or } Y = 22.04 + 0.649\ X$$

If X is 50, Y will be: $Y = 22.04 + 0.649\ (50) = 22.04 + 32.45 = 54.49$

Thus the estimated marks in Economics for a candidate who has obtained 50 marks in Statics are 54.49,

Example 25:

You are given below the following information about advertising and sales:

	Adv. Exp. (X) (Rs. crores)	*Sales (Y) (Rs. crores)*
Mean	*20*	*100*
S.D.	*5*	*12*

Correlation coefficient = 0.8

(a) Calculate the two regression lines.

(b) Find the likely sales when advertisement expenditure is Rs. 25 crores.

(c) What should be advertisement expenditure if the company wants to attain sales target of Rs. 130 crores?

Solution:

Regression equation of X on Y : $\overline{X} = r\frac{\sigma_x}{\sigma_y}(Y - \overline{Y})$

$\overline{X} = 20,\ \overline{Y} = 100,\ r = .8,\ \sigma_x = 5,\ \sigma_y = 12$

$X - 20 = .8\ \frac{5}{12}\ (Y - 100)$

$X - 20 = .333\ (Y - 100)$

$X - 20 = .333\ Y - 33.3 \quad \text{or} \quad X = -13.3 + 0.333\ Y$

Put $Y = 130$

$= .333\ (130) - 13.3 = 43.29 - 13.3 = 29.99$

$Y - \overline{Y} = r\frac{\sigma_y}{\sigma_x}(X - \overline{X})$

$Y - 100 = .8\ \frac{12}{5}\ (X - 20)$

$Y - 100 = 1.92\ X - 38.4 \quad \text{or}\ Y = 61.6 + 1.92\ X$

When $X = 25$, Y will be $Y = 1.92\ (25) + 61.6 = 109.6$.

Example 26:

Following is the distribution of students according to their height and weight Height in inches

	Weight in ibs.			
	90–100	*100–110*	*110–120*	*120–130*
50–55	*4*	*7*	*5*	*2*
55–60	*6*	*10*	*7*	*4*
60–65	*6*	*12*	*10*	*7*
65–70	*3*	*8*	*6*	*3*

Calculate (i) the coefficients of regression, and (ii) obtain the two regression equations.

Solution:

Let height be denoted by X and weight by Y.

Computation Required for Finding the Regression Equations

Y	X m d_y	d_x	90–100 95 – 2	100–110 105 –1	110–120 115 0	120–130 125 1	f	$f d_x$	$f d_y^2$	$f d_x d_y$
50–55	52.5	– 1	+8 4	+7 7	0 5	–2 2	18	–18	18	+13
55–60	57.5	0	0 6	0 10	0 7	0 4	27	0	0	0
60–65	62.5	1	–12 6	–12 12	0 10	+7 7	35	35	35	–17
65–70	67.5	2	–12 3	–16 8	0 6	+6 3	20	40	80	–22
		f	19	37	28	16	N = 100	Σfd_y = 57	Σfd_x^2 = 133	Σfd_xd_y = –26
		$f d_y$	– 38	–37	0	16	Σfd_y = –49			
		$f d_y^2$	76	37	0	16	Σfd_y^2 = 129			
		$f d_xd_y$	–16	–21	0	11	Σfd_xd_y = –26			

Regression coefficient of Y on X : $b_{yx} = \dfrac{N\Sigma fd_xd_y - \Sigma fd_x \, \Sigma fd_x}{N\Sigma fd_x^2 - (\Sigma fd_x)^2} \times \dfrac{i_y}{i_x}$

$\Sigma fd_xd_y = -26$, $\Sigma fd_x = 57$, $\Sigma fd_y = -59$, $N = 100$,

$\Sigma fd_x^2 = 133$, $i_x = 5$, $i_y = 10$

$$\frac{(100)\,(-26) - (57)\,(-59)}{(100)\,(133) - (57)^2} \times \frac{10}{5} = \frac{-2600 + 3363}{13300 - 3249} \times = 0.152$$

Regression coefficient of X on Y:

$$b_{xy} = \frac{N\Sigma fd_xd_y \Sigma fd_x \Sigma fd_y}{N\Sigma fd_y^2 - (\Sigma fd_y)^2} \times \frac{i_y}{i_x} = \frac{(100)\,(-26) - (57)\,(-59)}{(100)\,(129) - (-59)^2} \times \frac{5}{10}$$

$$= \frac{-2600 + 3363}{1200 - 3181} \times 0.5 = 0.041$$

Regression Equation of Y on X ; $Y - \overline{Y} = b_{yx}\,(X - \overline{X})$

$$\overline{Y} = A + \frac{\Sigma fd_y}{N} \times i = \frac{11559}{100} \times 10 = 109.1$$

$$\overline{X} = A + \frac{\Sigma fd_x}{N} \times i = 57.5 + \frac{57}{100} \times 5 = 57.5 + 2.85 = 60.35$$

$$Y - 109.1 = 0.152\ (X - 60.35) = 0.152\ X - 917$$

or $\qquad Y = 99.93 + 0.152$

Regression Equation of X on Y : $X - \overline{X} = b_{xy}\ (Y - \overline{Y})$

$$X - 60.35 = 0.041\ (Y - 109.1) = 0.041\ Y - 447$$

or $\qquad X = 55.88 + 0.041\ Y.$

Example 27:

The following talbe gives the aptitude test scores and productivity indices of 10 workers selected at random:

Aptitude scores (X)	*60*	*62*	*65*	*70*	*72*	*48*	*53*	*73*	*65*	*82*
Productivity index (Y)	*68*	*60*	*62*	*80*	*85*	*40*	*52*	*62*	*60*	*81*

Calculate the two regression equations and estimate (i) the productivity index of a worker whose test score is 92. (ii) The test score of a warder whose productivity index is 75.

Solution:

Calculation of Regression Equations

Aptitude Score X	**(X – 65) x**	**x^2**	**Productivity index Y**	**(Y – 65) y**	**y^2**	**xy**
60	–5	25	68	+3	9	–15
62	–3	9	60	–5	25	+15
65	0	0	62	–3	9	0
70	+5	25	80	+15	225	+75
72	+7	49	85	+20	400	+140
48	–17	289	40	–25	625	–425
53	–12	144	52	–13	169	+156
73	+8	64	62	–3	9	–24
65	0	0	60	–5	25	0
82	+17	289	81	+16	256	+272
$\Sigma X = 650$	$\Sigma x = 0$	$\Sigma x^2 = 894$	$\Sigma Y = 650$	$\Sigma y = 0$	$\Sigma y^2 = 1752$	$\Sigma xy = 1044$

Regression equation of X on Y, $X - \overline{X} = r\frac{\sigma_x}{\sigma_y}(Y - \overline{Y})$

$$r\frac{\sigma_x}{\sigma_y} = \frac{\Sigma_{xy}}{\sigma_{y^2}} = \frac{1044}{1752} = +\,0.596$$

$$\overline{X} = \frac{\Sigma X}{N} = \frac{650}{10} = 65;\ \overline{Y} = \frac{\Sigma Y}{N} = \frac{650}{10} = 65$$

$$X - 65 = .596\,(Y - 65)$$

$$X - 65 = .596\,Y - 38.74$$

or $$X = 26.26 + 0.596\,Y$$

For finding out the test score of a person whose productivity index is 75, put Y = 75 in the above equation:

$$X_{75} = 26.26 + 0.596\,(75) = 26.26 + 44.7 = 70.96.$$

Regression Equation of Y on X: Y $\overline{Y} = r\frac{\sigma_y}{\sigma_x}(X - \overline{X})$

$$r\frac{\sigma_y}{\sigma_x} = \frac{\Sigma xy}{\Sigma x^2} = \frac{1044}{894} = +\,1.168$$

$$Y - 65 = 1.168\,(X - 65)$$

$$Y - 65 = 1.168\,X - 75.92 \text{ or } Y = -\,10.92 + 1.168\,Y$$

For finding out the productivity index of a worker whose test score is 92 we put x = 92 in the above equation

$$-10.92 + 1.168\,(92)$$

$$= -\,10.92 + 107.456 = 96.536.$$

Example 28:

Find out the regression coefficients of Y on X and Y on Y from the following data:

$\Sigma X = 50,\ \overline{X} = 5,\ \Sigma Y = 60,$

$\overline{Y} = 6,\ \Sigma XY = 350.$

Variance of X = 4, Variance of Y = 9.

Solution:

Regression coefficient of Y on X;

$$b_{yx} = \frac{N\Sigma XY - \Sigma X \Sigma Y}{N\Sigma X^2 - (\Sigma X)^2}$$

W are not given the values of N, ΣX^2

$$\overline{X} = \frac{\Sigma X}{N} \text{ or } N\overline{X} = \Sigma E \text{ or } N \times 5 = 50 \text{ or } N = 10$$

ΣX^2 can be calculated from the formula of variance, i.e.,

$$\sigma_x^2 = \frac{\Sigma(X - \overline{X})^2}{N} = \frac{N(X - \overline{X})^2}{10}$$

$$\Sigma (X - \overline{X})^2 = 40$$

$$\text{or } \Sigma X^2 - N\,\overline{X}^2 = 40$$

$$\Sigma X^2 = 40 + 10\,(5)^2$$

$$\Sigma X^2 = 40 + 250 = 290$$

$$b_{yx} = \frac{(10 \times 350) - (50 \times 60)}{(10 \times 290) - (50)^2} = \frac{3500 - 3000}{2900 - 2500} = \frac{500}{400} = 1.25$$

Regression Equation of X on Y: $b_{xy} = \dfrac{N\Sigma XY - \Sigma X \Sigma Y}{N\Sigma Y^2 - (\Sigma Y)^2}$

ΣY^2 can be calculated as follows $\sigma_y^2 = \dfrac{\Sigma(Y - \overline{Y})^2}{N}$

$$10 \times 9 = \Sigma(Y - \overline{Y})^2$$

$$\Sigma Y^2 - N\overline{Y}^2 = 90 + 10\,(6)^2 = 450$$

$$b_{yx} = \frac{(10 \times 350) - (50 \times 60)}{(10 \times 450) - (60)^2} = \frac{3500 - 3000}{4500 - 3600} = \frac{500}{900} = 0.556$$

Example 29:

Height of the fathers and sons are given below. Find height of the son when the height of the father is 70 inches.

Father (inches) :	*71*	*68*	*66*	*67*	*70*	*71*	*70*	*73*	*72*	*65*	*66*
Son (inches):	*69*	*64*	*65*	*63*	*65*	*62*	*65*	*64*	*66*	*59*	*62*

Solution:

Denote father by X and son by Y. We have to fit regression of Y on X for answering for desired question.

Calculation of Regression line of Y on X

X	$(X-\overline{X})$			$(Y-\overline{Y})$		
	x	x^2	Y	y	y^2	xy
71	–2	4	69	+5	25	+10
68	–1	1	64	0	0	0
66	–3	9	65	+1	1	–3
67	–2	4	63	–1	1	+2
70	+1	1	65	+1	1	+1
71	+2	4	62	–2	4	–4
70	+1	1	65	+1	1	+1
73	+4	16	64	0	0	0
72	+3	9	66	+2	4	+6
65	–4	16	59	–5	25	+20
66	3	9	62	–2	4	+6
$\Sigma X = 759$	$\Sigma x = 0$	$\Sigma x^2 = 74$	$\Sigma Y = 704$	$\Sigma y = 0$	$\Sigma y^2 = 66$	$\Sigma xy = 39$

$$(Y-\overline{Y}) = r\frac{\sigma_y}{\sigma_x}(X-\overline{X})$$

$$\overline{Y} = \frac{704}{11} = 64,\ \overline{X} = \frac{759}{11} = 69$$

Since deviations are taken from actual means of X and Y:

$$r\frac{\sigma_y}{\sigma_x} = \frac{\Sigma xy}{\Sigma x^2} = \frac{39}{74} = 0.527$$

$$Y - 64 = .527\ (X - 69)$$

$$Y - 64 = .527\ X - 36.36 \text{ or } Y = .527\ X + 27.64$$

For X = 70, Y shall be: 527 (70) + 27.64 = 36.89 + 27.64 = 64.53

Hence the likely height of son when height of father is 70 inches shall be 64.53 inches.

Example 30:

From the data given below find:

(a) the two regression equations.

(b) the coefficient of correlation between marks in Economics and Statistics.

Marks in Economics :25 28 35 32 31 36 29 38 34 32

Marks in Statistics : 43 46 49 41 36 32 31 30 33 39

Solution:

Let marks in Economics be denoted by X and Statistics by Y.

Calculation of Regression Equations

X	(X – 32) x	x^2	Y	(Y – 38) y	y^2	xy
25	–7	49	43	+5	25	+35
28	–4	16	46	+8	64	–32
35	+3	9	49	+11	121	+33
32	0	0	41	+3	9	0
31	–1	1	36	–2	4	+2
36	+4	16	32	–6	36	–24
29	–3	9	31	–7	49	+21
38	+6	36	30	–8	64	–48
34	+2	4	33	–5	25	–10
32	0	0	39	+1	1	0
$\Sigma X = 320$	$\Sigma x = 0$	$\Sigma x^2 = 140$	$\Sigma Y = 380$	$\Sigma y = 0$	$\Sigma y^2 = 398$	$\Sigma xy = 93$

Regression equation of X on Y: $X - \overline{X} = r\frac{\sigma_x}{\sigma_y}(Y - \overline{Y})$

$$\overline{X} = \frac{\Sigma X}{N} = \frac{320}{10} = 32;\ \overline{Y} = \frac{\Sigma Y}{N} = \frac{380}{10} = 38$$

$$r\frac{\sigma_x}{\sigma_y} = \frac{\Sigma xy}{\Sigma y^2} = \frac{-93}{398} = -0.234$$

$$X - 32 = -.234\ (Y - 38)$$

$$X - 32 = -.234\ Y + 8.892$$

$$X = -.234\ Y + 40.892 \quad \text{or} \quad X = 40.892 - .234\ Y$$

Regression equation of Y on X : $Y - \overline{Y} = r\frac{\sigma_y}{\sigma_x}(X - \overline{X})$

$$r\frac{\sigma_y}{\sigma_x} = \frac{\Sigma xy}{\Sigma x^2} = \frac{-93}{140} = -0.664$$

$$Y - 38 = -0.664\ (X - 32)$$

$$Y - 38 = -.664X + 21.248 \quad \text{or} \quad Y = 59.248 - 0.664X.$$

Coefficient of Correlation

$$r = \sqrt{b_{xy} \times b_{yx}} = \sqrt{-.234 \times -664} = -0.394$$

Likely marks in Statistics when marks in Economics are 30

$$Y = 59.248 - .664X$$

Putting $X = 30$

$$Y = 59.248 - .664(30) = 59.248 - 19.92 = 39.328 \text{ or } 39.$$

Example 31:

The equations of two lines of regression obtained in a correlation analysis the following:

$$2x = 8 - 3Y \text{ and } 2Y = 5 - X$$

Obtain the value of the correlation coefficient.

Solution:

Let regression eq. $2X = 8 - 3Y$ be the regression of X on Y:

$$2X = 8 - 3Y$$

$$Y = 4 - \frac{3}{2}Y$$

$$b_{xy} = -\frac{3}{2}$$

From eq. (ii) $\quad Y = \frac{5}{2} - \frac{1}{2}X$ or $b_{yx} = -\frac{1}{2}$

$$r = \sqrt{b_{xy} \times b_{yx}} = -\sqrt{\frac{3}{2} \times \frac{1}{2}} = 0.866$$

Since r is less than 1, our assumption is correct.

Example 32:

A student obtained the following regression equations do you agree with him;

$$6X = 15Y + 21 \qquad \text{...(1)}$$

$$21X + 14Y = 56 \qquad \text{...(ii)}$$

Solution:

Treating eq. (i) as regression eq. of X on Y : $6X = 15Y + 21$

$$X = \frac{15}{6}Y + \frac{21}{6}. \text{ Hence } b_{xy} = \frac{15}{6}$$

From eq. (ii) 14Y = 56 – 21X

$$Y = \frac{56}{14} - \frac{21}{14}X$$

Here $b_{yx} = \frac{56}{14}$

Since both the regression coefficients are greater then one it is not possible. Reversing the assumption and taking eq (i) as regression of Y on X:

$$15Y = 6X - 21$$

$$Y = \frac{6}{15}X - \frac{21}{15} \text{ or } b_{yx} = \frac{6}{15}$$

From eq. (ii) $21X = 56 - 14Y$

$$X = \frac{56}{21} - \frac{14}{21}Y \text{ or } b_{xy} = -\frac{14}{21}$$

Since it is not possible that one regression coefficient is negative and another positive, the equations are wrongly obtained by the student.

Example 33:

Given the following data:

$$\overline{X} = 36, \ \overline{Y} = 85, \ \sigma_x = 11, \ \sigma_y = 8, \ r = 0.66$$

Find the two regression equations and estimate the value of X when Y = 75.

Solution:

Regression equation of X on Y : $X - \overline{X} = r\frac{\sigma_x}{\sigma_y}(Y - \overline{Y})$

$$X - 36 = .66\ \frac{11}{8}\ (Y - 85)$$

$$X = 36 = 0.9075\ (Y - 85)$$

$$X - 36 = 0.9075\ Y - 77.1375 \text{ or } X = -41.1375 + 0.9075Y$$

Estimated value of X when Y is 75 shall be:

$$X = -4.11375 + 0.9075\ (75) = 26.925$$

Regression equation of Y on X : $Y - \overline{Y} = r\frac{\sigma_y}{\sigma_x}(X - \overline{X})$

$$Y - 85 = 66\frac{8}{11}(X - 36)$$

$$Y - 85 = 0.48\ (X - 36)$$

$$Y - 85 = 0.48\ X - 17.28 \text{ or } Y = 67.72 + 0.48\ X.$$

Example 34:

Calculate the correlation coefficient from the following results:

$$N = 10,\ \Sigma X = 350,\ \Sigma Y = 310$$

$$\Sigma(X - 35)^2 = 162,\ \Sigma(Y - 31)^2 = 222,$$

$$\Sigma(X - 35)\ (Y - 31) = 92.$$

Also find the regression line and Y on X

Solution:

We are given $n = 10$, $\Sigma X = 350$, $\Sigma Y = 310$

$$\overline{X} = \frac{350}{10} = 35,\ \overline{Y} = \frac{310}{10} = 31$$

$$r = \frac{\Sigma(X - \overline{X})(Y - \overline{Y})}{\sqrt{\Sigma(X - \overline{X})^2}\sqrt{\Sigma(Y - \overline{Y})^2}}$$

$$\Sigma(X - \overline{X})(Y - \overline{Y}) = 92,\ \Sigma(X - \overline{X})^2 = 162,\ \Sigma(Y - \overline{Y})^2 = 222$$

$$r = \frac{92}{\sqrt{162 \times 222}} = \frac{92}{189.64} = 0.485$$

Regression line of Y on X : $= Y - \overline{Y} = b_{yx}(X - \overline{X})$

$$b_{yx} = r\frac{\sigma_y}{\sigma_x} = \frac{\Sigma xy}{\Sigma x^2} = \frac{92}{162} = 0.568$$

$$Y - 31 = .568\ (X - 35)$$

$$Y - 31 = .568\ X - 19.88 \text{ or } Y = .568\ X + 11.12$$

Example 35:

If two regression coefficient are 0.8 and 0.6 what would be the value of the coefficient of correlation?

Solution:

Coefficient of correlation is the under-root of the two regression coefficients:

i.e., $$r = \sqrt{b_{xy} \times b_{yx}} = \sqrt{0.8 \times 0.6} = 0.693$$

Example 36:

You are given below the following information about advertisement expenditure and sales:

	Adv. Exp. (X) (RS. crores)	*Sales (Y) (Rs. crores)*
Mean	*20*	*120*
S.D.	*5*	*25*

Correlation coefficient 0.8

(i) Calculate the two regression equations.

(ii) Find the likely sales when advertisement expenditure is Rs. 25 crores.

(iii) What should be the advertisement budget if the company wants to attain sales target of Rs. 150 crores?

Solution:

(i) Regression equation of x on Y : $X - \overline{X} = r\frac{\sigma_x}{\sigma_y}(Y - \overline{Y})$

$$\overline{X} = 20,\ \overline{Y} = 120;\ \sigma_x = 5,\ \sigma_y = 25,\ r = 0.8$$

$$X - 20 = .8\ \frac{8}{25}\ (Y - 120)$$

$$X - 20 = 0.16\ (Y - 120)$$

$$X - 20 = 0.16\ Y - 19.2 \quad \text{or } X = 0.8 + 0.16\ Y$$

Regression equation of Y on X : $Y - \overline{Y} = r\frac{\sigma_y}{\sigma_x}(X - \overline{X})$

$$Y - 120 = .8\ \frac{25}{5}\ (X - 20)$$

$$Y - 120 = 4\ (X - 20)$$

$$Y - 120 = 4\ X - 80 \quad \text{or} \quad Y = 40 + 4\ X$$

(ii) From regression equation of Y on X, we can find out the likely sales when advertisement expenditure is Rs. 25 crores $Y_{25} = 4\ (25) + 40 = 140$

Hence the likely sales when advertisement expenditure is Rs. 25 crores is Rs. 140 corores.

(iii) From regression equation of X on Y, we can find the advertisement budget if the Cowants to attain sales target of Rs. 150 crores.

$$X_{150} = .16\ (150) + 0.8 = 24 + 8 = 24.8.$$

Hence to attain a sales target of Rs. 150 crores, the advertisement budget should be Rs. 24.8 crores.

Example 37:

Calculate the regression equation of X on Y and Y on X from the following data and estimate X when Y = 26:

X:	*10*	*12*	*13*	*17*	*18*	*20*	*24*	*30*
Y:	*5*	*6*	*7*	*9*	*13*	*15*	*20*	*21*

Calculate coefficient of correlation also.

Solution:

Calculation of Regression Equations

X	**(X – 18)**		**Y**	**(Y – 12)**		
	x	**x^2**		**y**	**y^2**	**xy**
10	–8	64	5	–7	49	56
12	–6	36	6	–6	36	36
13	–5	25	7	–5	25	25
17	–1	1	9	–3	9	3
18	0	0	13	1	1	0
20	+2	4	15	+3	9	6
24	+6	36	20	+8	64	48
30	+12	144	21	+9	81	108
$\Sigma X = 144$	$\Sigma x = 0$	$\Sigma x^2 = 310$	$\Sigma Y = 96$	$\Sigma y = 0$	$\Sigma y^2 = 274$	$\Sigma xy = 282$

Regression equation of X on Y : $X - \overline{X} = b_{xy} = (Y - \overline{Y})$

$$b_{xy} = \frac{\Sigma xy}{\Sigma y^2} = \frac{282}{274} = 1.029$$

$X - 18 = 1.029\ (Y - 12)$

$X - 18 = 1.294\ Y - 12.\ 248$ or $X = 5.652 + 1.0294\ Y$

Estimated value of X when Y = 26: $x = 1.029\ (26) + 5.652 = 32.406$

Regression equation of Y on X : $Y - \overline{Y}\ b_{yx}\ (X - \overline{X})$

$$b_{yx} = \frac{\Sigma xy}{\Sigma x^2} = \frac{282}{310} = 0.91$$

$Y - 12 = .91\ (X - 18)$

$Y - 12 = 91\ X - 16.38$ or $Y = -4.38 + 0.91\ X$

$$r = \sqrt{b_{xy} \times b_{yx}} = \sqrt{1.029 \times 91} = 0.968$$

Example 38:

Estimate (a) the sale for advertising expenditure of Rs. 100 lakhs and (b) the advertisement expenditure for sales of Rs. 47 crores from the data given below:

Sales (Rs. crores):	*14*	*15*	*18*	*20*	*24*	*30*	*32*
Adv. Exp. (Rs. lakhs):	*52*	*62*	*65*	*70*	*76*	*80*	*78*

Solution:

Since sales depend on advisement expenditure, let sales be denoted by Y and advertisement expenditure by X.

We have to find two regression equations for answering two given questions;

Calculation for Regression Equations

Sales Y	**(Y – 22)** y	y^2	**Adv Exp. X**	**(X – 69)** x	x^2	**xy**
14	–8	64	52	–17	289	+136
16	–6	36	62	–7	49	+42
18	–4	16	65	–4	16	+16
20	–2	4	70	+1	1	–2
24	+2	4	76	+7	49	+14
30	+8	64	80	+11	121	+88
32	+10	100	78	+9	81	+90
$\Sigma X = 154$	$\Sigma y = 0$	$\Sigma y^2 = 288$	$\Sigma x = 483$	$\Sigma x = 0$	$\Sigma x^2 = 606$	$\Sigma xy = 384$

Regression equation of X on Y:

$$X - \overline{X} = r\frac{\sigma_x}{\sigma_y}(Y - \overline{Y})$$

$$\overline{X} = \frac{\Sigma X}{N} = \frac{483}{7} = 69;\ \overline{Y} = \frac{\Sigma Y}{N} = \frac{154}{7} = 22$$

Since deviations or taken from actual means

$$r\frac{\sigma_x}{\sigma_y} = \frac{\Sigma xy}{\Sigma y^2} = \frac{384}{288} = 1.333$$

$$X - 69 = 1.333\ (Y - 22)$$

$$X - 69 = 1.333\ Y - 29.326 \quad \text{or} \quad X = 1.333\ Y + 39.674$$

when Y = 47, X shall be:

1.333 (47) + 39.674 = 62.651 + 39.674 = 102.325

For attaining sales target of 47 crores we require an advertisement expenditure of Rs. 102.325 crores.

Regression equation of Y on X:

$$Y - \overline{Y} = r\frac{\sigma_y}{\sigma_x}(X - \overline{X})$$

$$\overline{Y} = \frac{\Sigma Y}{N} = \frac{154}{7} = 22,$$

$$\overline{X} = \frac{\Sigma X}{N} = \frac{483}{7} = 69$$

Since deviations are taken from actual means

$$r\frac{\sigma_y}{\sigma_x} = \frac{\Sigma xy}{\Sigma x^2} = \frac{384}{606} = 0.634$$

$$Y - 22 = .634\ (X - 69)$$

$$Y - 22 = .634\ X - 43.75$$

$$Y = .634\ X - 21.75$$

For X = 100, Y shall be .634 (100) – 21.75 = 41.65

Thus the likely sales for advertisement ercenditure of Rs. 100 lakhs is Rs. 41.65 crores,

Example 39:

Comment on the following:

For a bivariate distribution:

(i) $b_{xy} = 2.8$ *and* $b_{yx} = -0.3$

(ii) $b_{xy} = -0.8$, $b_{yx} = -1.2$ *and* $r = 0.92$.

Solution:

(i) We are given $b_{xy} = 2.8$ and $b_{yx} = -0.3$. One of the properties of regression coefficients is that both of them will have the same sign. Since one regression coefficient is positive and another negative, there is something wrong with the given information.

(ii) We are given $b_{xy} = -0.8$ and $b_{yx} = -1.2$, $r = 0.98$.

The coefficient of correlation is the under-root of the two regression coefficient and it shall have the same sign as the regression coefficients. The value of r should be

$$-\sqrt{8 \times 1.2} = -0.98$$

We are given r = 0.98 whereas it should be – 0.98.

Example 40:

From 10 observations of price (m) and supply (n) of a commodity, the following summary figures were obtained (in appropriate units):

$$\Sigma m = 130;\ \Sigma n = 220;\ \Sigma m^2 = 2288;\ \Sigma n^2 = 5506 \text{ and } \Sigma mn = 3467$$

Compute a line of regression of n on m and estimate the supply when the price is 16

Solution:

Regression line of n on m is given by :

$$n = \overline{n} = r\frac{\sigma_n}{\sigma_m}(m - \overline{m})$$

$$\overline{n} = \frac{\Sigma n}{n} = \frac{220}{10} = 22;\ \overline{m} = \frac{\Sigma m}{n} = \frac{130}{10} = 13$$

$$r\frac{\sigma_n}{\sigma_m} = \frac{\Sigma mn - \frac{\Sigma n \times \Sigma m}{n}}{\Sigma m^2 - \frac{(\Sigma m)^2}{n}}$$

$$= \frac{3467 - \frac{220 \times 130}{10}}{2288 - \frac{(130)^2}{10}} = \frac{3467 - 2860}{2288 - 1690} = \frac{607}{598} = 1.015$$

$$n - 22 = 1.015\ (m - 13)$$

$$n - 22 = 1.015\ m - 13.195$$

$n = 1.015\ m + 8.805$ when m = 16, n shall be:

$$n = 1.015\ (16) + 8.805 = 16.24 + 8.805 = 25.045$$

Thus the estimated supply when price is 16 units is 25.045.

Example 41:

The correlation coefficient between two variables X and Y is r = 0.6. If $S_x = 1.50$, $s_y = 2.00$, $\overline{X} = 10$ and $\overline{Y} = 20$, find the regression lines of:

(i) Y on X and (ii) X on Y

Solution:

Regression line of X on Y:

$$X - \overline{X} = r\frac{S_x}{S_y}(Y - \overline{Y})$$

$$\overline{X} = 10,\ r = 06,\ S_x = 1.5,\ S_y = 2,\ \overline{Y} = 20$$

$$X - 10 = .6\frac{1.5}{2}\ (y - 20)$$

$$X - 10 = .45\ (X - 20)$$

$$X - 10 = .45\ Y - 9 \quad \text{or } X = 1 + 0.45\ Y$$

Regression line of Y on X, $Y - \overline{Y} = r\frac{S_y}{S_x}(X - \overline{X})$

$$Y - 20 = .6\ \frac{2}{1.5}\ (X - 10)$$

$$Y - 20 = .8\ (X - 10)$$

$$Y - 20 = 0.8\ X\ Y = 12 + 0.8\ X.$$

Example 42:

Find two regression equations for the following two series, what is most likely value of X when Y = 20 and Most likely value of Y when x = 22.

X: 35	*25*	*29*	*31*	*27*	*24*	*33*	*36*
Y: 23	*27*	*26*	*21*	*24*	*20*	*29*	*30*

Solution:

Calculation of Regression Equations

X	**$(X - \overline{X})$** **x**	**x^2**	**Y**	**$(Y - \overline{Y})$** **y**	**y^2**	**xy**
35	+5	25	23	–2	4	–10
25	–5	25	27	+2	4	–10
29	–1	1	26	+1	1	–1
31	+1	1	21	–4	16	–4
27	–3	9	24	–1	1	+3
24	–6	36	20	–5	27	+30
33	+3	9	29	+4	16	+12
36	+6	36	30	+5	25	+30
$\Sigma X = 240$	$\Sigma x = 0$	$\Sigma x^2 = 142$	$\Sigma Y = 200$	$\Sigma y = 0$	$\Sigma y^2 = 92$	$\Sigma xy = 50$

Regression equation of X on Y : $X - \overline{X} = r\frac{\sigma_x}{\sigma_y}(Y - \overline{Y})$

$$r\frac{\sigma_x}{\sigma_y} = \frac{\Sigma xy}{\Sigma y^2} = \frac{30}{92}$$

$$\overline{X} = \frac{\Sigma X}{N} = \frac{240}{8} = 30;\ \overline{Y} = \frac{\Sigma Y}{N} = \frac{200}{8} = 25$$

$$X - 30 = 0.544\ (Y - 25)$$

$$X - 30 = 543\ Y - 13.575$$

or $X = 16.425 + 0.543\ Y$

$$X_{20} = 16.425 + 0.543\ (20) = 27.285$$

Regression equation of Y on X: $Y - \overline{Y} = r\frac{\sigma_y}{\sigma_x}(X - \overline{X})$

$$r\frac{\sigma_y}{\sigma_x} = \frac{\Sigma xy}{\Sigma x^2} = \frac{50}{142} = 0.352$$

$$X - 25 = 0.352\ (X - 30)$$

$$Y - 25 = 0.352\ X - 10.684$$

or $Y = 14.316 + 0.352\ X$

$$Y_{22} = 14.316 + 0.352\ (22) = 14.316 + 7.744 = 22.06$$

Example 43:

A panel of judges A and B graded seven debaters and independently awarded the following marks:

Debaters	*1*	*2*	*3*	*4*	*5*	*6*	*7*
Marks by A	*40*	*34*	*28*	*30*	*44*	*38*	*31*
Marks by B	*32*	*39*	*26*	*30*	*38*	*34*	*28*

An eighth debater was awarded 36 marks by judge A while judge B was not present.

If judge B were also present, how many marks would you expect him to award to the eighth debator assuming that the same degree of relationship exists in their judgment?

Solution:

Let marks by A be denoted by X and marks by B by Y. The marks expected to be awarded by judge B can be determined by fitting regression equation of Y on X.

Marks by A X	(X – 35) dx	dx^2	Marks by B Y	(Y – 30) d_y	d_y^2	$d_x d_y$
40	+5	25	32	+2	4	+10
34	–1	1	39	+9	81	–9
28	–7	49	26	–4	16	+28
30	–5	25	30	0	0	0
44	+9	81	38	+8	64	+72
38	+3	9	34	+4	16	+12
31	–4	16	28	–2	4	+8
$\Sigma X = 245$	$\Sigma d_x=0$	$\Sigma d_x^2 = 206$	$\Sigma Y = 227$	$\Sigma d_y = 17$	$\Sigma d_y^2 = 182$	$\Sigma d_x d_y=121$

Regression equation of Y on X : $Y - \overline{Y} = r\frac{\sigma_y}{\sigma_x}(X - \overline{X})$

$$\overline{Y} = \frac{\Sigma Y}{N} = \frac{227}{7} = 32.43;\ \overline{X} = \frac{\Sigma X}{N} = \frac{245}{7} = 35$$

$$r\frac{\sigma_y}{\sigma_x} = \frac{\Sigma d_x d_y - \frac{\Sigma d_x \times \Sigma d_y}{N}}{\Sigma d_x^2 - \frac{(\Sigma d_y)^2}{N}} = \frac{121 - \frac{0 \times 17}{7}}{206 - \frac{(0)^2}{7}} = \frac{121}{206} = 0.587$$

$$Y - 32.43 = .587\ (X - 35)$$

$$Y - 32.43 = .587\ X - 20.545$$

or $Y = 11.885 + 0.587\ Y$

For $X = 36$

$$Y = .587\ (36) + 11.885 = 21.132 + 11.885 = 33.017 \text{ or } 33.$$

Thus if judge B were also present he would have awarded 33.017 or 33 marks to the eighth debator.

Example 44:

Two regression lines of a sample are X + 6Y = 6 and 3x + 2Y = 0. Find the correlation coefficient.

Solution:

We are give two equations:

$$X + 6Y = 6 \qquad ...(i)$$

$$3X + 2Y = 10 \quad \text{...(ii)}$$

Taking eq. (i) as the regression of Y on X : 6Y = 6 – X

$$Y = 1 - \frac{1}{6}X \quad \text{or } b_{yx} = -\frac{1}{6}$$

From eq. (ii) $3X = 10 - 2Y$

$$X = \frac{10}{3} - \frac{2}{3}Y \text{ or } b_{xy} = -\frac{2}{3}$$

$$r = \sqrt{b_{xy} \times b_{yx}} = -\sqrt{\frac{1}{6} \times \frac{2}{3}} = -0.333$$

Example 45:

For a given set of bivariate data, the following results were obtained.

$\overline{X} = 53.2$, $\overline{Y} = 27.9$, *Regression coefficient of Y on X = – 1.5.*

Regression coefficient of X on Y = – 0.2. Find the most probable value of Y when X is 60.

Solution:

For finding the most probable value of Y when X = 60, we will fit a regression eq. of Y on X.

$$Y - \overline{Y} = r\frac{\sigma_y}{\sigma_x}(X - \overline{X})$$

$$\overline{Y} = 27.9,\ r\frac{\sigma_y}{\sigma_x} = -1.5,\ \overline{X} = 53.2$$

$$Y - 27.0 = -1.5\ (X - 53.2)$$

$$Y - 27.9 = -1.5\ x - 79.8$$

$$\text{or } Y = 107.7 - 1.5\ X$$

$$Y_{60} = 107.7 - 1.5\ (60) = 17.7$$

$$r = \sqrt{b_{xy} \times b_{yx}} = -\sqrt{0.2 \times 1.5} = -0.548$$

Example 46:

The following data relate to marks in Advanced Accounts and Business Statistics in B. Com. (H), II yr Examination of a particular year in Delhi university:

Mean Marks in Advanced Accounts	*= 30*
Mean Marks in Business Statistics	*= 5*

Standard Deviation of Marks in Advanced Accounts = *10*

Standard Deviation of Marks in Business Statistics = *7*

Coefficient of Correlation between the Marks of

Advanced Accounts and Business Statics = *0.8*

Form the two regression lines and calculate the expected marks in Advanced Accounts if, the marks secured by a student in Business Statistics are 40.

Solution:

Let marks in Statistics be debator by Y and Accountancy by X.

Regression line of X on Y : $X - \overline{X} = r\frac{\sigma_x}{\sigma_y}(Y - \overline{Y})$

$$\overline{X} = 30,\ \overline{Y} = 35,\ \sigma_x = 10,\ \sigma_y,\ r = 0.8$$

$$X - 30 = .8\ \frac{10}{7}\ (Y - 35)$$

$$X - 30 = 1.134\ (Y - 35)$$

$$X - 30 = 1.143\ Y - 40.005$$

or $$X = -\ 10.005 + 1.143\ Y$$

Regression line of Y on X : $Y - \overline{Y} = r\frac{\sigma_y}{\sigma_x}(X - \overline{X})$

$$Y - 35 = .8\ \frac{7}{10}\ (X - 30)$$

$$Y - 35 = .56\ (X - 30)$$

$$Y - 35 = .56 - 16.8$$

or $$Y = 18.2 + 0.56\ X$$

The marks in advanced accounts for a student who has got 40 marks in business statistics can be obtained from regression equation of

$$X = 1.143\ (40) - 10.005 = 45.72 - 10.005 = 35.715 \text{ or } 36.$$

Example 47:

By using the following data, find out the two lines of regression and from them compute the Kari Pearson's coefficient of correlation:

$\Sigma X = 250;\ \Sigma XY = 7{,}900;$

$\Sigma X^2 = 6{,}500;$

$\Sigma Y^2 = 10{,}000$ *and* $N = 10.$

Solution:

Regression line of X on Y : $X - \overline{X} = r\frac{\sigma_x}{\sigma_y}(Y - \overline{Y})$

$$\overline{X} = \frac{\Sigma X}{N} = \frac{250}{10} = 25\ ;\ \overline{Y} = \frac{\Sigma Y}{N} = \frac{300}{10} = 30$$

$$r\frac{\sigma_x}{\sigma_y} = \frac{\Sigma XY - N\overline{X}\,\overline{Y}}{\Sigma Y^2 - N(\overline{Y})^2}$$

$$= \frac{7900 - (10 \times 25 \times 30)}{10000 - 10(30)^2} = \frac{7900 - 7500}{10000 - 9000} = .4$$

$$X - 25 = .4\ (Y - 30)$$

$$X - 25 = .4\ Y - 12 \quad \text{or} \quad x = 13 + 0.4\ Y$$

Regression line of Y on X : $Y - \overline{Y} = r\frac{\sigma_y}{\sigma_x}(X - \overline{X})$

$$r\frac{\sigma_y}{\sigma_x} = \frac{\Sigma XY - N\overline{X}\,\overline{Y}}{\Sigma X^2 - N(\overline{X})^2}$$

$$= \frac{7900 - (10 \times 25 \times 30)}{6500 - 10(25)^2} = \frac{7900 - 7400}{6500 - 6250} = 1.6$$

$$Y - 30 = 1.6\ (X - 25)$$

$$Y - 30 = 1.6\ X - 40 \quad \text{or} \quad Y = 1.6\ X - 10$$

Karl Perarson's correlation coefficient:

$$r = \sqrt{b_{xy} \times b_{yx}} = \sqrt{.4 \times 1.6} = 0.8$$

LIMITATIONS OF REGRESSION ANALYSIS

In making estimate from a regression equation, it is important to remember that the assumption is being made that relationship has not changed since the regression equation was computed. Another point worth remembering is that the relationship shown by the scatter diagram may not be the same if the equation is extended beyond the values used in computing the equation. For example, there may be a close linear relationship between the yield of a crop and the amount of fertilizer applied. with the yield increasing as the amount of fertilzer is increased. It would not be logical, however, to extend this equation beyond the limits of the experiment for it is quite likely that if the amount of fertilizer were increased indefinitely, the yield would eventually decline as too much fertilizer was applied.

LIST OF FORMULAES

Regression equation of X on Y: $X - \overline{X} = r\frac{\sigma_x}{\sigma_y}(Y - \overline{Y})$

$$r\frac{\sigma_x}{\sigma_y} = \frac{\Sigma xy}{\Sigma y^2}$$

(if deviations are taken from actual means of X and Y)

$$r\frac{\sigma_x}{\sigma_y} = \frac{\Sigma d_x \Sigma d_y - \frac{\Sigma d_x \times \Sigma d_x}{N}}{\Sigma d_y^2 - \frac{(\Sigma d_y)^2}{N}} \quad \text{Like for } r\frac{\sqrt{y}}{\sqrt{x}}$$

(if deviations are taken from assumed means of X and Y)

Regression equation of Y on X: $Y - \overline{Y} = r\frac{\sigma_y}{\sigma_x}(X - \overline{X})$

$$r\frac{\sigma_y}{\sigma_x} = \frac{\Sigma xy}{\Sigma x^2} \text{ (Deviations taken from actual mean)}$$

$$r\frac{\sigma_y}{\sigma_x} = \frac{\Sigma d_x d_y - \frac{\Sigma d_x - \Sigma d_y}{N}}{\Sigma d_y^2 - \frac{(\Sigma d_x)^2}{N}}$$

(Deviations taken from assumed mean)

Regression coefficients $r\frac{\sigma_X}{\sigma_y}$ or b_{yx} is the regression coefficient of X on Y.

$r\frac{\sigma_y}{\sigma_x}$ or b_{yx} is the regression coefficient of Y on X.

$$r = \sqrt{b_{xy} \times b_{yx}}$$

Standard errors of estimate $S_{xy} = \pm\sqrt{\frac{\Sigma(X - X_c)^2}{N}}$ also $S_{xy} = \sigma_x\sqrt{1 - r^2}$

$$S_{xy} = \sqrt{\frac{\Sigma(Y - Y_c)^2}{N}} \text{ also } S_{yx} = \sigma_y\sqrt{1 - r^2}$$

EXERCISES

1. A survey was conducted to study the relationship between expenditure on accommodation (X) and expenditure on food and entertainment (Y) and the following results were obtained :

	Mean	*S.D.*
Expenditure on accommodation :	Rs. 173	63.15
Expenditure on food and entertainment :	Rs. 47.8	22.98

Coefficient of correlation = + 0.57

Write down the equation of regression on X and Y and estimate the expenditure on food and entertainment, if the expenditure on accommodation is Rs. 200.

[$Y = 0.207X + 11.99$; $Y_{200} = 53.29$]

2. (a) For a bivariate data X = 20, Y = 25, $s_x = 5$, $s_y = 8$ and r = 0.7, find the regression equation of X on Y and estimate the value of X when Y is 30.

[$X = 0.43754 + 9.062$: $X_{30} = 22.187$]

(b) Given the two regression coefficients, X on Y, 1.5 and Y on X, 0.2, find the correlation coefficient.

[r = 0.548]

3. For a bivariate data, you are given the following information :

$$\Sigma (X - 58) = 46$$
$$\Sigma (X - 58)^2 = 3086$$
$$\Sigma (Y - 58) = 9$$
$$\Sigma (Y - 58)^2 = 483$$
$$[\Sigma (X - 58) (Y - 58) = 1095]$$

Number of pairs of observation = 7.

You are required to determine (i) the two regression equations and (ii) the coefficient of correlation between X and Y series.

[$Y = 35.399 + 0.37X$; $X = -65.87 + 2.24$, $r = +0.902$]

4. (a) What are regression coefficients? Point out their important properties.

(b) Let b_{yx} and b_{xy} stand for the coefficients of regression of Y on X and X on Y respectively. Show that :

$$r_{xy} = \sqrt{b_{xy} \times b_{yx}}.$$

5. (a) Point out the role of regression analysis in business decision-making. What are the important properties of regression coefficients?
   (b) Explain the concept of lines of regression. State why there are two lines of regression.
6. How do you fit a regression equation to a set of bivariate data? Explain. How is correlation analysis different from regression analysis?
7. (a) Clearly bring out the distinction between correlation and regression.
   (b) What is meant by a simple regression model ? Under what assumptions are the parameters of the model estimated ?
   (c) Explain the principal of least squares used for estimation, for linear regression.
8. (a) What are regression lines ? With the help of an example, illustrate how these lines are helpful in decision-making.
   (b) Explain why there are two regression equations. Under what conditions can there be one regression equation ?
9. (a) Distinguish clearly between 'Correlation' and 'Regression'. Why would there be two lines of regression in a bivariate distribution?
   (b) Define the standard error of the estimate of Y as obtained from the line of regression of Y on X. What does it represent?
10. From the following two regression equations, calculate the mean values of X and Y and the coefficient of correlation between X and Y :

$$4Y - 5X = 0$$

$$5Y - X - 63 = 0.$$

11. The following data about the sales and raw material expenses for a woman's self employment generation programme run by a NGO is given below :

	Sales (Rs. '000)	*Raw material expenditure (Rs. '000)*
$\overline{X}$	40	6
σ	10	15
r	+ 0.9	

Find the two regression equations.

12. In a regression study, the two regression lines are obtained as $2x - 3y + 6 = 0$ and $4y - 5x - 8 = 0$. Calculate the means of X and Y. If the standard deviation of X is 3, determine the standard deviation of Y and also the standard error in estimating Y from X.

(a) Find the two regression equations from the following data :

X :	1	2	3	4
Y :	8	6	4	2

13. Find the regression equation from the following data :

Age of husband :	18	19	20	21	22	23	24	25	26	27
Age of wife :	17	17	18	18	19	19	19	20	21	22

Also calculate the correlation coefficient between the ages of husbands and wives.

14. (a) Explain the term regression and state the difference between regression and correlation.

(b) What is regression? Why are there, in general, two regression lines? Under what conditions can there be only one regression line?

15. (a) What are regression coefficients? How are they computed?

(b) What do you understand by regression analysis? What are its advantages and disadvantages?

(c) What is meant by regression analysis? How does it help in business decision making?

16. (a) Explain the concept of regression and ratio of variation and state their utility in economic analysis.

(b) Comment on the following results obtained from given data :

(i) Coefficient of regression of Y on X is 4.2 and

(ii) Coefficient of regression of X on Y is 0.5.

17. The price of (X) and demand (Y) of a commodity are given below:

X :	10	12	14	16	18	20
Y :	80	72	60	51	45	38

(i) Estimate a linear demand function y = a + bx and find the demand when the price is 25.

(ii) Estimate the price elasticly of demand.

18. (a) Given the following data, compute the two regression coefficients and Karl Pearson's coefficient of correlation :

X/Y	0 – 20	20 – 40	40 – 60
10 – 25	10	5	3
25 – 40	4	40	8
40 – 55	6	9	15

[b_{xy} = 0.40, b_{yx} = 0.27, r = + 0.36]

(b) If X = 15 Y = 12, $\Sigma XY = 150$, $S_x = 6.4$, $S_y = 9.0$, N = 10, find the following :

(a) Regression coefficient of Y on X.

(b) Regression coefficient of X on Y, and

(c) Correlation coefficient between X and Y.

19. Fill in the blanks :

(i) The statistical tool, with the help of which we are in a position to estimate the ..... of one variable from the ...... values of another variable, is called ......

(ii) The regression analysis helps us to study the ...... relationship between the variables.

(iii) If both the regression coefficients are negative, the correlation coefficient would be ....

(iv) The under-root of two ..... coefficients gives us the value of correlation coefficient.

(v) The variable, we are trying to predict, is called the.

(vi) Both the regression coefficients cannot .... one.

[**Ans.** (i) unknown values, known regression (ii) nature (iii) negative (iv) regression (v) dependent variable (vi) exceed]

20. Find the most likely price in Bombay corresponding to the price of Rs. 70 at Calcutta from the following data :

	Calcutta	*Bombay*
Average Price :	65	67
Standard Deviation :	2.5	3

Correlation coeff. between the prices of the commodities in the two cities is 0.8.

[$Y = 5.8 + 1.12X$: $Y_{70} = 72.6$ taking Calcutta as X and Bombay Y]

21. The following results of capital employed and profit earned by a firm in 10 successive years are calculated :

	Mean	*Standard deviation*	*Coefficient correlation*
Capital employed : (*Rs. thousands)*	Rs. 55	Rs. 28.7	+ 0.96
Profit eared : *(Rs. thousands)*	Rs. 13	Rs. 85	

(i) Obtain the two regression equations.

(ii) Estimate the amount of profits to be earned if capital employed is Rs. 50,000.

(iii) Estimate the amount of capital to be employed if profit earned is Rs. 20,000.

[(i) X = 12.867 + 3.241Y: Y = 2.62 + 0.284X, (ii) 14202.62 (iii) 64832.67]

22. What do you understand by regression? What are the properties of regression coefficients ?

23. (a) Distinguish between regression and correlation. State the formula for the product moment correlation coefficient (r) and explain the significance of r = + 1 and r = – 1.

(b) Prove that the regression lines of X on Y and Y on X intersect at the point [X, Y].

24. (a) Explain the difference between correlation and regression analysis. (B. Com., Mysore Univ., 1998)

(b) What is regression line? What are its uses? Why are there generally two regression lines?

(c) Briefly explain the concept of regression and write down the equations of the regression lines? When do the lines coincide?

25. In trying to evaluate the effectiveness of its advertising campaign, a firm completed the following information :

Year :	1991	1992	1993	1994	1995	1996	1997	1998
Adv. Expenditure (Rs. '000) :	12	15	15	23	24	38	42	48
Sales (lak Rs.) :	5.0	5.6	5.8	7.0	7.2	8.8	9.2	9.5

Calculate the regression equation of sales on advertising expenditure is Rs. 60.000.

26. (a) In a regression study the two regression lines are obtained as 2x – 3y + 6 = 0 and 4y – 5x – 80. Calculate means of X and Y. If the standard deviation of x is 3 determine the standard deviation of Y.

(b) The two regression equations are 3x – 2y + 1 = 0 and 3x – 8y + 13 = 0. Find the arithmetic means X and Y.

27. The following tables gives the age of cars of a certain make and annual maintenance costs. Obtain the regression equation for costs related to age.

Age of cars (in years) :	2	4	6	8
Maintenance costs (in hundreds of Rs.)	10	20	25	30

[X = 4.75 + 3.25Y]

28. Calculate the coefficient of correlation and two lines of regression from the following data :

Sales Revenue (Lakh Rs.)	*Advertising Expenditure*			
	5–15	15–25	25–30	35–45
75–125	3	4	4	8
125–175	8	6	5	7
175–225	2	2	3	4
225–275	3	3	2	2

[X = 172.655 – 0.5165Y; Y = 30.4 – 0.0273X; r = 0.119]

29. Obtain the equations of the two lines of regression for the following data:

X :	43	44	46	40	44	42	45	42	38	40	42	57
Y :	29	31	19	18	19	27	27	29	41	30	26	10

Hence, obtain the correlation coefficient between X and Y. For what value of X, Y = 49?

30. Tick the correct answer :

(a) The regression lines cut each other at the point of :

(i) average of X and Y (ii) average of X only (iii) average of Y only (iv) the median of X and Y (v) none of these

(b) The regression line of Y on X :

(i) minimises total of the square of the horizontal deviations (ii) total of the squares of the vertical deviations (iii) both vertical and horizontal deviations (iv) none of them

(c) There will be only one regression line in case of two variables, if : (i) r = 0 (ii) r = + 1 (iii) r = – 1 (iv) r is either + 1 or (v) r is very low.

(d) The farther the two regression lines cut each other :

(i) greater will be the degree of correlation. (ii) the lesser will be the degree of correlation (iii) does not matter (iv) none of these.

(e) Where r is zero the regression lines cut each other making an angle of (i) 30 (ii) 60 (iii) 90 i.e. parallel to OY and OX (iv) none of these.

(f) the regression line of Y on X passes through the plotted points in such a manner that :

(i) $\Sigma (Y - Yc)^2 = 0$ (ii) $\Sigma (Y - Xc) = 0$ (iii) $\Sigma (Y - Yc) = 0$ (iv) $\Sigma (X - Yc) = 0$ (v) $\Sigma (Yc - Y)^2 = 0$.

(g) The greater the value of r :

(i) the better are estimates, obtain through regression analysis (ii) the worst are the estimates (ii) really makes no difference, (iv) none of these.

(h) When one regression coefficient is negative, the other would be:

(i) bivariate (ii) positive (iii) zero (iv) none of these

(i) With $b_{xy} = 0.5$ and $r = 0.8$ and variance of $Y = 16$ the standard deviation of X equal to ....

(i) 2.5 (ii) 6.4 (iii) 10.0 (iv) 25.6.

[**Ans.** (a) (i), (b) (i), (c) (iv), (d) (ii), (e) (iv), (f) (iii), (g) (i), (h) (i)]

31. From the following data obtain the two regression equations :

Sales :	91	97	108	121	67	124	51	73	111	57
Purchases :	71	75	69	97	70	91	39	61	80	47

Also find the correlation coefficient between the sales and purchases.

32. (a) Find the coefficient of correlation if regression coefficient of X on Y is 0.84 and regression coefficient of Y on X is 0.4. Are the above coefficients correct?

[r – 0.5797]

(b) The coefficient of correlation between ages of husbands and wives in a community was found to be + 0.8, the average of husband's age was 25 years and that of wives age 22 years. Their standard deviations were, respectively, 4 and 5. Find with the help of regression equations :

(i) The expected age of husband when wife's age is 20 years, and

(ii) the expected age of wife when husband's age is 33 years.

[$X = .64Y + 10.92$, $X_{20} = 23.72$, $Y = X - 3$, $Y_{33} = 30$].

33. The following table gives the ages of husbands and wives for 50 newly married couples. Find the two regression lines. Estimate the age of husband when the age of wife is 20 and the age of wife when the age of husband is 30.

Age of wives	*Age of husbands*			*Total*
	20–25	25–30	30–35	
16–20	9	14	–	23
20–24	6	11	3	20
24–28	–	–	7	7
Total	15	25	10	50

[$Y = 8.27 + 0.904X$; $X = 26.35$, $Y = 21.85$]

24. Find the two regression equations from the following data :

X :	6	2	10	5	9
Y :	9	11	5	9	8

35. A student obtained the following two regression equations. Do you agree with him?

[6x = 15y + 2], 21x + 15y = 56.

36. You are given the following data :

Variable	X	Y
Mean	47	96
Variance	64	81

Coefficient of correlation between X and Y = 0.36. Determine the equations of regression lines. Calculate Y when X = 50 and X when Y = 88.

[X = 0.32Y + 6.28, Y = .405X + 76.965, Y_{50} = 97.215, X_{88} = 44.44].

37. From the following information, calculate line of regression of Y on X :

	X	*Y*
Mean :	40	60
Standard Deviation (SD) :	10	15
Correlation Coefficient :	(r)	= 0.7

38. A financial manager speculates about the relationship between family incomes and their allocation for investment. The following table presents the results of a survey of 8 randomly selected families :

Annual Income (in '000 Rs.)	8	12	9	24	13	37	10	16
Per cent allocation for Investment :	36	25	33	15	28	19	20	22

(a) Develop a regression equation that best describes this data.

(b) Estimate the per cent allocation for investment of family earning Rs. 25,000 annually.

[(a) Y = 0.576X + 14.814 (b) 29.214].

39. In an examination of relationship between yield of wheat and rainfall, the following results were obtained :

	Yield	*Rainfall*
Mean	900 kg.	12 inches
S.D.	80 kg.	2 inches

Coefficient of correlation = + 0.5

Calculate the likely yield when rainfall is 15 inches.

40. In a partially destroyed laboratory record of an analysis of correlation data, the following results are only legible :

Variance of X = 9, Regression equations :

$$4X - 5Y + 33 = 0$$

$$20X - 9Y' - 107 = 0$$

Find (i) The mean values of X on Y (ii) The standard deviation of Y (iii) The correlation coefficient.

[$\overline{X} = 13$, $\overline{Y} = 17$, $r = 0.6$, $s_y = 0.667$].

41. (a) Given the means of two variables, X and Y are 68 and 150, their standard deviations are 2.5 and 20 and coefficient of correlation between them is 0.6, write down the regression equations of X on Y.

(b) If the two regression lines are $3Y - 9X = 46$ and $3X + 12Y = 19$, determine which one of these is the regression in line of Y on X and which one is that of X on Y. Also find the means correlation coefficient and ratio of the variances of X and Y.

42. For a sample of firms, the correlation between the profit rate (X) and growth rate (Y) is found to be + 0.9. Given the following additional information :

(i) compute the regression equation of Y on X, and

(ii) estimate the likely growth rate when the profit rate is 16%.

	Profit Rate (%)	*Growth Rate (%)*
Mean :	20	25
Standard Deviation :	3	5

43. The following data about the sales and advertisement expenditure of a firm is given below :

	Sales	*Advertisement expenditure (Rs. crores)*
Mean	40	6
S.D.	10	1.5

Coefficient of correlation, r = 0.9.

(i) Estimate the likely sales for a proposed advertisement expenditure of Rs. 10 crores.

(ii) What should be advertisement expenditure if the firm proposes a sales target of 60 crores of rupees?

[(i) 64 crores (iii) 8.7 crores].

44. The age and blood pressure of 10 women are :

Age :	56	42	36	47	49	42	60	72	63	55
Blood pressure :	147	125	118	128	145	140	155	164	149	150

(a) Find the correlation coefficient between age and blood pressure.

(b) Determine the least square regression equation of blood pressure on age.

(c) Estimate the blood pressure of a woman whose age is 45 years.

45. Given the following information :

	X Rs.	*Y* Rs.
Arithmetic average :	6	8
Standard deviation :	5	40/3

Coefficient of correlation between X and Y = 8/15.

Find :

(i) the regression equation of Y on X.

(ii) the regression equation of X on Y, and

(iii) the most likely value of Y when X = 100 rupees.

[X = 0.2Y + 4.4: Y = 1.42X + 2.82; Y = 141.48].

46. (a) Find out the regression equation showing the regression of capacity utilization (Y) on production (X) from the following data. The correlation coefficient between X and Y is 0.62.

	Mean	*Standard deviation*
Production (in lakh units)	35.6	10.5
Capacity utilization (in percentage)	84.8	8.5

Estimate the production when the capacity utilization is 70%.

(b) If the regression equations of y on x and x on y are respectively y = 2x and 6x – y = 4 and the second moment of x about the origin is 3. Find (i) the correlation coefficient, and (ii) standard deviation of y.

47. Construct two regression equations for the following data :

X :	23	43	53	63	73	83
Y :	5	6	7	8	9	10

[Y = 0.857X + 26.73; X = 11.428Y – 29.38].

48. Derive regression lines for the following data :

S X = 30, S X^2 = 190,

S XY = 192; S Y = 30,

S $Y^2 = 190$, n = 5

[Y = 1.011X – 0.066; X = 0.914Y + 0.516].

49. Given that y = mx + 4 and x = 4y + 5 the regression lines of y on x and x on y, respectively, show that $0 < m \leq 0.25$. If actually m 0.10, find the means of x and y and also their correlation coefficient. Can you find their standard deviation as well?

50. (a) What do you mean by linear regression? Obtain the equation for the line of regression of Y on X.
    (b) Explain the concept of regression and point out its importance in business forecasting.

51. (a) What is regression? How is it different from correlation?
    (b) What are the assumptions of least square regression method? What problems are faced when these are violated?
    (c) What is regression analysis? Discuss its utility in predicting future events.
    (d) Explain the concept of regression and point out its usefulness in dealing with the given problems.
    (e) Define 'regression' Distinguish between regression and correlation.

52. (a) Explain the concept of standard error of estimate of the linear regression of Y on X. Can you express it in terms of correlation coefficient? What is the standard error of estimating Y from X if R = 1?
    (b) Explain the concept of regression. How does it differ from correlation?
    (c) What is meant when it is said that the smaller the standard error of estimate, the better the regression line 'fits' the data? How is the standard error of estimate related to the possible usefulness of a regression line?

53. Given the following data, estimate (i) The value of 'Y' when 'X' = 70. (ii) The value of 'X' when 'Y' = 90.

	X	*Y*
Mean	18	100
Standard deviation	14	20

Coefficient of correlation = + 0.8.

54. (a) Which of the following statements is True or False :
    (i) Regression analysis reveals average relationship between two variables. T/F

(ii) The regression lines cut each other at the point of average of X and Y. T/F

(iii) Regression coefficient of Y on X measures the change in X corresponding to a unit change in Y. T/F

(iv) Regression coefficients are independent of change of scale and origin. T/F

(v) If standard error of estimate is zero, there is no variation about the regression line an correlation will be perfect. T/F

(vi) The term 'regression' was first used by Karl Pearson in the year 1990. T/F

(vii) The regression coefficient of Y on X is denoted by the symbol b_{xy}. T/F

(viii) The term 'dependent' and 'independent' do not imply that there is necessarily any cause-effect relationship between the variables. T/F

[**Ans.** (i) T (ii) T (iii) F (iv) F (v) T (vi) F (vii) F (viii) T]

3

Partial and Multiple Correlation

INTRODUCTION

The basic distinction between multiple and partial correlation analysis is that whereas in the former we measure the degree of relationship between the variable Y and all the variables. X_1, X_2, X_3 .... X_n, taken together, in the latter we measure the degree of relationship between Y and one of the variables X_1, X_2, X_3 ..... X_n with te effect of all the other variables removed.

PARTIAL CORRELATION

It is often important to measure the correlation between a dependent variable and one particular independent variable when all other variables involved are kept constant *i.e.*. when the effects of all other variables are removed (often indicated by the phrase "other things being equal"). This can be obtained by calculating coefficient of partial correlation. Thus partial correlation analysis measures the strength of the relationship between Y and one independent variable in such a way that variations in te other independent variables are taken into account. A partial correlation coefficient is analogous to a partial regression coefficient in that all other factors are "held constant". Simple correlation, on the other hand, ignores the effect of all other variables even though these variables might be quite closely related to the independent variable, on to one another.

Zero Order, First Order and Second Order Coefficients

Partial coefficients such as $r_{12.3}$, $r_{13.2}$ are often referred to as first order coefficients, since one variable has been held constant. Simple coefficients (correlation between two variables only) are called zero order coefficients, since no variables are held constant. $r_{12.34}$, $r_{13.24}$, etc., are called second order coefficients since two variables are kept constant. Stated generally, the order of designation indicates the number of variables that have been held constant statistically.

Partial Correlation Coefficient

Partial correlation coefficient provides a measure of the relationship between the dependent variable and other variables, with the effect of the most of the variables eliminated.

If we denote by r_{123} the coefficient of partial correlation between X_1 and X_2 keeping X_3 constant, we find that

$$r_{12.3} = \frac{r_{12} - r_{13}\ r_{23}}{\sqrt{(1-r_{13}{}^2)}\sqrt{(1-r_{23}{}^2)}}$$

Similarly,

$$r_{13.2} = \frac{r_{13} - r_{12}\ r_{23}}{\sqrt{(1-r^2{}_{12})}\sqrt{(1-r^2{}_{23})}}$$

where $r_{13.2}$ is the coefficient of partial correlation between X_1 and X_2 keeping X_2 constant.

$$r_{23.1} = \frac{r_{23} - r_{12}\ r_{13}}{\sqrt{(1-r^2{}_{12})}\sqrt{(1-r^2\,13)}}$$

where $r_{23.1}$ is the coefficient of partial correlation between X_2 and X_3 keeping X1 constant.

Thus, for three variables, X_1, X_2 and X_3, there will be three coefficients of partial correlation each studying the relationship between two variables when the third is held constant.

The partial correlation coefficient helps us to answer questions such as this. Is the correlation between, say, X_1 and X_2, merely due to the fact that both are affected by X_2 or is there a no covariation between X_1 and X_2 over and above the association due to the common influence of X_3? Thus, in determining a partial correlation coefficient between X_1 and X_2, we attempt to remove the influence of X_3 from each of the two variables so as to ascertain whether net relationship exists between the "unexplained" residuals that remain.

Example 1:

Is it possible to get the following from a set of experimental data:

(a) $r_{23} = 0.8\ r_{31} = -0.5,\ r_{12} = 0.6$

(b) $r_{23} = 0.7,\ r_{31} = -0.4,\ r_{12} = 0.6.$

Solution (a):

$$r_{12.3} = \frac{r_{12} - r_{13}r_{23}}{\sqrt{\left(1-r^2{}_{13}\right)}\sqrt{\left(1-r^2\,23\right)}}$$

$$= \frac{0.6-(-.05)\,(.8)}{\sqrt{0.75}\sqrt{0.36}} = \frac{1}{0.52} = 1.923$$

Since the value of $r_{12.3}$ is greater than one, there is some inconsistency in the given data.

(b)
$$r_{12.3} = \frac{r_{12} - r_{13}r_{23}}{\sqrt{\left(1-r^2{}_{13}\right)}\sqrt{\left(1-r^2{}_{23}\right)}}$$

$$= \frac{0.6-(-0.4)(0.7)}{\sqrt{1-(-0.4)^2}\sqrt{1-(0.7)^2}}$$

$$= \frac{0.6+0.28}{\sqrt{0.84}\sqrt{0.51}} = \frac{0.88}{0.655} = 1.344\,.$$

This again is greater than one which is not possible.

Example 2:

On te basis of observations made on 39 cotton plants, the total correlation of yield of cotton (X_1), number of bolls, i.e., seed vessels (X_2) and height (X_3) and height (X_3) are fod to be:

$$r_{12} = 0.8,\ r_{13} = 0.65 \text{ and } r_{23} = 0.7$$

Commenting the partial correlation between yield of cotton and the number of bolls, eliminating the effect of height.

Solution:

We have to find the partial correlation between yield of cotton and te number of bolls, eliminating the effect of height, *i.e.*, in terms of symbols, we have to calculate $r_{12.3}$.

$$r_{12.3} = \frac{r_{12} - r_{13}\ r_{23}}{\sqrt{(1-r^2 13)}\sqrt{(1-r^2 23)}}$$

$$r_{12} = 0.8,\ r_{13} = 0.65 \text{ and } r_{23} = 0.7$$

Substituting the values

$$r_{12.3} = \frac{0.8-(0.65\times 0.7)}{\sqrt{1-(0.65)^2}\sqrt{1-(0.7)^2}}$$

$$= \frac{0.8-0.455}{\sqrt{1-.4225}\sqrt{1-0.49}}$$

$$= \frac{0.345}{0.76 \times 0.714} = \frac{0.345}{0.543} = 0.635$$

Example 3:

If $r_{12} = 0.86$, $r_{13} = 0.65$ *and* $r_{23} = 0.72$, *find the partial correlation coefficient* $r_{12.3}$.

Solution:

$$r_{12.3} = \frac{r_{12} - r_{13}\ r_{23}}{\sqrt{(1.r^2 13)}\sqrt{(1-r^2 23)}}$$

$$r_{12} = 0.86,\ r_{13} = 0.65,\ r_{23} = 0.72$$

Substituting the values

$$r_{12.3} = \frac{.86 - .65 \times .72}{\sqrt{1-(.65)^2}\ \sqrt{1-(.72)^2}}$$

$$= \frac{.86 - .468}{\sqrt{.5775 \times 4816}} = \frac{.392}{.5274} = 0.743$$

Example 4:

On the basis of the following information compute:

(i) $r_{23.1}$ *(ii)* $r_{13.2}$ *and (iii)* $r_{12.3}$:

$r_{12} = 0.70;\ r_{13} = 0.61;\ r_{23} = 0.40$

Solution:

$$r_{23.1} = \frac{r_{23} - r_{12}\ r_{13}}{\sqrt{(1-r^2 12)}\sqrt{(1-r^2 13)}}$$

Substituting te given values

$$r_{23.1} = \frac{0.4 - 0.7 \times 0.61}{\sqrt{1-(0.7)^2}\ \sqrt{1-(0.61)^2}}$$

$$= \frac{0.4 - 0.427}{\sqrt{0.51}\sqrt{1-0.3721}} = \frac{-0.027}{0.714 \times 0.792} = 0.048$$

$$r_{13.2} = \frac{r_{13} - r_{12}\ r_{23}}{\sqrt{(1-r^2 12)}\sqrt{(1-r^2 23)}}$$

$$= \frac{0.61-(0.7)(0.4)}{\sqrt{1-(0.7)^2}\sqrt{1-(0.4)^2}} = \frac{0.61-0.28}{\sqrt{1-0.49}\sqrt{1-0.16}}$$

$$= \frac{0.33}{\sqrt{0.51}\sqrt{0.84}} = 0.504$$

$$r_{12.3} = \frac{r_{12} - r_{13}\, r_{23}}{\sqrt{1-r^2 13}\sqrt{1-r^2 23}}$$

$$= \frac{0.7-(0.61\times 0.4)}{\sqrt{1-(0.61)^2}\sqrt{1-(0.4)^2}}$$

$$= \frac{0.7-0.244}{\sqrt{1-0.3721}\sqrt{1-0.16}} = 0.629$$

Second-order Partial Correlation Coefficients

Second order coefficients may be obtained from first order coefficients. In case of four variables. If $r_{12.34}$ is the coefficient of partial correlation between X_1 and X_2 keeping X_3 and X_4 constant, then

$$r_{12.34} = \frac{r_{12.4} - r_{13.4}\, r_{23.4}}{\sqrt{(1-r^2{}_{13.4})}\sqrt{(1-r^2{}_{23.4})}}$$

Similarly $$r_{13.24} = \frac{r_{13.4} - r_{12.4}\, r_{23.4}}{\sqrt{(1-r^2{}_{12.4})}\sqrt{(1-r^2{}_{23.4})}}$$

and $$r_{14.23} = \frac{r_{14.3} - r_{12.3}\, r_{24.3}}{\sqrt{(1-r^2{}_{12.3})}\sqrt{(1-r^2{}_{24.3})}}$$

Alternative formulae giving the same results are available for all three of the second-order coefficients. They are:

$$r_{12.34} = \frac{r_{12.3} - r_{14.3}\, r_{24.3}}{\sqrt{(1-r^2{}_{14.3})}\sqrt{(1-r^2{}_{24.3})}}$$

$$r_{13.24} = \frac{r_{13.2} - r_{14.2}\, r_{34.2}}{\sqrt{(1-r^2{}_{14.2})}\sqrt{(1-r^2{}_{34.2})}}$$

PARTIAL CORRELATION COEFFICIENTS IN CASE OF FOUR VARIABLES

When four variables are involved in a correlation problem, there are twelve possible first-order coefficients. Some of these are:

$$r_{14.2} = \frac{r_{14} - r_{12}\, r_{24}}{\sqrt{\left(1 - r^2_{12}\right)}\sqrt{\left(1 - {}^2_{24}\right)}}$$

$$r_{14.3} = \frac{r_{14} - r_{13}\, r_{34}}{\sqrt{\left(1 - r^2_{13}\right)}\sqrt{\left(1 - {}^2_{34}\right)}}$$

$$r_{13.4} = \frac{r_{13} - r_{14}\, r_{34}}{\sqrt{\left(1 - r^2_{14}\right)}\sqrt{\left(1 - {}^2_{34}\right)}}$$

$$r_{12.4} = \frac{r_{12} - r_{14}\, r_{24}}{\sqrt{\left(1 - r^2_{14}\right)}\sqrt{\left(1 - {}^2_{24}\right)}}$$

$$r_{24.3} = \frac{r_{24} - r_{23}\, r_{34}}{\sqrt{\left(1 - r^2_{23}\right)}\sqrt{\left(1 - {}^2_{34}\right)}}$$

$$r_{34.2} = \frac{r_{34} - r_{23}\, r_{24}}{\sqrt{\left(1 - r^2_{23}\right)}\sqrt{\left(1 - {}^2_{24}\right)}}$$

$$r_{23.4} = \frac{r_{23} - r_{24}\, r_{34}}{\sqrt{\left(1 - r^2_{24}\right)}\sqrt{\left(1 - {}^2_{34}\right)}}$$

In a similar manner, the formulae for other partial correlation coefficients, *i.e.* $r_{12.3}$, $r_{13.2}$, $r_{24.1}$, $r_{34.1}$ can also be written.

$$r_{14.23} = \frac{r_{14.2} - r_{13.2}\, r_{34.2}}{\sqrt{\left(1 - r^2_{12.3}\right)}\sqrt{\left(1 - {}^2_{34.1}\right)}}$$

The value of a partial correlation correlation coefficient is usually interpreted via the corresponding coefficient of partial determination, which is merely the square of the former. Thus if $r_{12.3} = 0.4$, $r^2_{12.3} = 0.16$.

The t-test employed to test the significance of a sample correlation can be employed to test the significance of a partial correlation when the number of degrees of freedom is reduced by the number of variables eliminate.

Characteristic and Uses of Partial Correlation Analysis

The function of partial correlation analysis is the measurement of relationship between two factors, with the effects of one or more other factors eliminated. If the assumptions of the method are true for a series of data, the power f partial analysis is great. The problem of holding certain variables constant, while the relationship between the others is measured, often presents difficulty itself in statistical analysis. Partial correlation is especially useful in the analysis of interrelated series. It is particularly pertinent to uncontrolled experiments of various kinds, in which such interrelationships usually exist. Most economic data fall in this category.

Partial correlation is of greatest value when used in conjunction with gross and multiple correlation in the analysis of factors affecting variations in many kinds of phenomena.

Partial analysis, like all correlation, has the advantage that the relationships are expressed concisely in a few well-defined coefficients. Also it is adaptable to small amounts of data and the reliability of the results can be rather easily tested.

Limitations of Partial Correlation Analysis

(1) The usefulness of the partial analysis is somewhat limited by the following basic assumptions of the method:

   (i) The gross or zero-order correlation must have linear regressions.

   (ii) The effects of the independent variables must be additively and not jointly related.

   (iii) Because the reliability of partial coefficients decreases as its order increases, the number of observations in gross correlations should be fairly large. After the student carries the analysis beyond the limits of the data. Thus, weakness to some extent can be guarded against by test of reliability.

(2) When the above assumptions have been satisfied, partial analysis still possesses the disadvantages of laborious calculations and difficult interpretation even for statisticians.

The interpretation of the partial and multiple correlation results tends to assume that the independent variables have causal effects on dependent variable. This assumption is sometimes true, but more often untrue in varying degrees.

The Significance of a Partial Correlation Coefficient

The significance of a partial r may be determined, most readily, by the *Z* transformation. The S.E. $= \frac{1}{\sqrt{N-3}}$ and the S.E. of the *Z* corresponding to $r_{12.3}$ is $= \frac{1}{\sqrt{N-4}}$. One degree of freedom is subtracted from *N* for each variable eliminated, in addition to the 3 already lost. So for $r_{12.34}$ the S.E. of the corresponding *Z* is $\frac{1}{\sqrt{N-3-2}} = \frac{1}{\sqrt{N-5}}$.

Example:

The following zero-order correlation coefficients are given

$$r_{12} = 0.98,\ r_{13} = 0.44 \text{ and } r_{23} = 0.54.$$

Calculate multiple correlation coefficient treating first variable as dependent and second and third variables as independent.

Solution:

We have to calculate the multiple correlation coefficient treating first variable as dependent and second and third variables as independent, *i.e.*, we have to find $R_{1.23}$.

$$R^2 = \frac{\text{Explained variation}}{\text{Total variation}}$$

$$R_{1.23} = \frac{\sqrt{r^2_{12} + r^2_{13} - 2r_{12}\, r_{13}\, r_{23}}}{1 - r^2_{23}}$$

Substituting the given values $R_{1.23}$

$$= \frac{\sqrt{(.98)^2 + .(44)^2 - 2(.98)(.44)(.54)}}{1-(.54)^2}$$

$$= \frac{\sqrt{.9604 + .1936 - .4657}}{.7084} = 0.986.$$

MULTIPLE CORRELATION

In problems of multiple correlation we are dealing with situations that involve three or more variables. For example, we may consider the association between the yield of wheat per acre and both the amount of rainfall and the average daily temperature. We are trying to make estimates of the value of one of the these variables based on the values of all the others. The variables whose value we are trying to estimate is called the dependent variable and the other variables on which our estimates are based are known as independent variables. The statistician himself chooses which variable is to be dependent and which variables are to be independent. It is merely a question of problem being studies. If we are trying to determine the most problem weight of men, we make weight of men, we make weight the dependent variable and height, age, etc., independent variables. If on the other hand, we are interested in estimating height, we will make height the dependent variable and weight, age, etc., the independent variables. Thus in problems of multiple correlation we always have three or more variables (one dependent and the other independent). In order that we may distinguish them easily we follow the custom of representing them by the letter X with subscript. The dependent variable is always denoted by X_1 and the other by X_2, X_3, etc. Thus in the height, age and weight problem, if we are trying to estimate men's weight (that is, if weight be a dependent variable), we might denote.

$X_1 \rightarrow$ weight in lbs.

$X_2 \rightarrow$ height in inches.

$X_3 \rightarrow$ age in years.

Coefficient of Multiple Correlation

the coefficient of multiple linear correlation is represented by R_1 and it is common to add subscripts designating the variables involved. Thus, $R_{1.234}$ would represent the coefficient of multiple linear correlation between X_1, on the one hand and X_2, X_3 and X_4 on the other. The subscript of the dependent variable is always to the left of the point.

The coefficient of multiple correlation can be expressed in terms of r_{12}, r_{13} and r_{23} as follows:

$$R_{1.23} = \frac{\sqrt{r^2_{12} + r^2_{13} - 2r_{12}\,r_{13}\,r_{23}}}{1 - r^2_{23}}$$

$$R_{2.13} = \frac{\sqrt{r^2_{12} + r^2_{23} - 2r_{12}\,r_{13}\,r_{23}}}{1 - r^2_{13}}$$

$$R_{3.12} = \frac{\sqrt{r^2_{13} + r^2_{23} - 2r_{12}\,r_{13}\,r_{23}}}{1 - r^2_{12}}$$

It should be noted that $R_{1.23}$.

A coefficient of multiple correlation such as $R_{1.23}$ lies between 0 and 1. The closer it is to 1, the better is the linear relationship between the variables. The closer it is to 0, the worse is the linear relationship. If the coefficient of multiple correlation is 1, the correlation is called perfect. Although a correlation coefficient of 0 indicates no linear relationship between the variables, it is possible that a non-linear relationship may exist. It should be noted that whereas the correlation coefficients range from + 1.0 to 0 to – 1.0, the coefficients of multiple correlation are always positive in sign and range from + 1.0 to 0. Since some of the individual variables may be positively correlated with the dependent variables and other negatively correlated, no useful purpose would be served in distinguishing between a positive and negative value of R.

By squaring $R_{1.23}$, we obtain te coefficient of multiple determination. An alternate formula for obtaining the value of $R_{1.23}$ is follows:

$$R_{1.23} = \sqrt{r^2_{12} + r_{13.2}\left(1 - r^2_{12}\right)}$$

or $$R^2_{1.12} = r^2_{12} + r_{13.2}\left(1 - r^2_{12}\right)$$

Similarly

$$R_{1.24} = \frac{\sqrt{r^2_{12} + r^2_{14} - 2r_{12}\,r_{14}\,r_{24}}}{1 - r^2_{24}}$$

or $$R_{1.24} = \sqrt{r^2_{12} + r^2_{14.2}\left(1 - r^2_{12}\right)}$$

and $$R_{1.34} = \frac{\sqrt{r^2_{13} + r^2_{14} - 2r_{13}r_{14}r_{34}}}{1 - r^2_{34}}$$

or $$R_{1.34} = \sqrt{r^2_{12} + r^2_{14.2}\left(1 - r^2_{13}\right)}$$

To determine a multiple coefficient with three independent variables, the following formula shall be used:

$$R_{1.234} = \sqrt{1 - \left(1 - r^2_{14}\right)\left(1 - r^2_{13.4}\right)\left(1 - r^2_{12.34}\right)}$$

Example:

Suppose that $r_{12.34} = 0.5$ and $N = 41$. Is the partial r significant? Also determine 95% confidence limits.

Solution:

$$\text{S.E.} = \frac{1}{\sqrt{N-5}} = \frac{1}{\sqrt{41-5}} = 0.167$$

The *Z* corresponding to *r* of 0.50 is 0.55. Hence $r_{12.34}$ is significant.

The 95% confidence interval for the population *Z* is

$0.55 \pm 1.96 \times 0.167$ or from 0.223 to 0.877.

Advantages of Multiple Correlation Analysis

The coefficient of multiple correlation serves the following purposes:

(1) It also serves as a measure of goodness of fit of the calculated plane of regression and consequently as a measure of the general degree of accuracy of estimates made by reference to equation for the plane of regression.

(2) It serves as a measure of the degree of association between one variable taken as the dependent variable and a group of other variables taken as the independent variables.

Limitations of Multiple Correlation Analysis.

(1) Linear multiple correlation involves a great deal of work relative to the results frequently obtained. When the result are obtained, only a few students, well-trained in the method are able to interpret them. The misuse of correlation results has probability led to more doubt on the method than is justified. However, tis lack of understanding and resulting misuse are due to the complexity of the method.

(2) Multiple correlation analysis is based on the assumption that the relationship between the variables is linear. In other words, the rate of change in one variable in terms of another is assumed to be constant for all values. In practice, most relationships are not linear but follow some other pattern. This limits somewhat the use of multiple correlation analysis. The linear regression coefficients are not accurately descriptive of curvilinear data.

(3) A second important limitation is te assumption that effects of independent variables on the dependent variables are separate, distance and additive. When the effects of variables are additive, a given change in one has the same effect on the dependent variable regardless of the sizes of the other two independent variables.

MULTIPLE REGRESSION ANALYSIS

Multiple Regression and Correlation Analysis

Multiple regression analysis represents a logical extension of two-variable regression analysis. Instead of a single independent variable, two or more independent variables are used to estimate the values of a dependent variable.

However, the fundamental concept in the analysis remains the same. The following are the three main objectives of multiple regression and correlation analysis:

(1) To derive an equation which provides estimates of the dependent variable from values of the two or more independent variables.

(2) To obtain a measure of the error involved in using this regression equation as a basis for estimation.

(3) To obtain a measure of the proportion of variance in the dependent variable accounted for or "explained by" the independent variables.

The first purpose is accomplished by deriving an appropriate regression equation by the method of least squares. The second purpose is achieved through the calculation of a standard error of estimate. The third purpose is accomplished by computing the multiple coefficient of determination.

MULTIPLE REGRESSION EQUATION

The multiple regression equation describes the average relationship between these variables and this relationship is used to predict or control the dependent variable.

A regression equation is an equation for estimating a dependent variable, say, X_1 from the independent variables X_2, X_3, ... and is called a regression equation of X_1 on X_2, X_3, ... In functional notation, this is sometimes written briefly as X1 = $F(X_2, X_3...)$ read as "X_1 is a function of X_2, X_3 and so on".

In case of three variables, the regression equation of X_1 on X_2 and X_3 has the form

$$X_{1.23} = a_{1.23} + b_{12.3}\ X_2 + b_{13.2}\ X_3 \qquad ...(i)$$

$X_{1.23}$ is the computed or estimated value of the dependent variable and X_2. X_3 are the independent variables.

The constant $a_{1.23}$ is the intercept made by the regression plane. It gives the value of the dependent variable when all the independent variables assume a value equal to zero. $b_{12.3}$ and $b_{13.2}$ are called partial regression coefficients or the net regression coefficients. $b_{12.3}$ measures the amount by which a unit change in X_2 is expected to affect X_1 when X_3 is held constant and $b_{13.2}$ measures the amount of change in X_1 per unit change in X_3 when X_2 is held constant.

Due to the fact the X_1 varies partially because of variation in X_2 and partially because of variation in X_3, we call $b_{12.2}$ and $b_{13.2}$ the partial regression coefficients of X_1 on X_2 keeping X_3 constant and of X_1 on X_3 keeping X_2 constant.

Normal Equation for the Least Square Regression Plane

Just as there exist least square regression lines approximating a set of N data points (X', Y) in a two - dimensional scatter diagram so also there exist least square regression planes fitting a set of N data points (X_1, X_2, X_3) in a three - dimensional scatter diagram.

The least square regression plane of X_1 on X_2 has the equation (i), where $b_{12.3}$ and $b_{13.2}$ are determined by solving simultaneously the normal equations

$$\Sigma X_1 = N a_{1.23} + b_{12.3} \Sigma X_2 + b_{13.2} \Sigma X_3$$

$$\Sigma X_1 X_2 = a_{1.23} \Sigma X_2 + b_{12.3} \Sigma X_2^2 + b_{13.2} \Sigma X_2 X_3$$

$$\Sigma X_1 X_3 = a_{1.23} \Sigma X_3 + b_{12.3} \Sigma X_2 X_3 + b_{13.2} \Sigma X_2^2$$

These equations can be obtained by multiplying both sides of equation (i) by 1, X_2 and X_3 successively and summing on both sides.

When the number of variables is 4 or more, solving the above system of normal equation becomes a very tedious procedure. Efficient methods of solving simultaneous equations require a knowledge of matrix algebra, which is not assumed for the reader of this text. Computer programmes are widely available for determining the variables of the constants in a multiple regression equation. In our discussion that follows, we shall confine ourselves to the two independent variables case which of course, can be extended to cover case with three or more independent variables.

Assumptions of Linear Multiple Regression Analysis

For point estimation, the principle assumptions of linear multiple regression analysis are:

(1) The dependent variable is a random variable whereas the independent variable need not be random variable.

(2) The relationship between the several independent variables and the one dependent variable is linear, and

(3) The variances of the conditional distributions of the dependent variable, given various combinations of values of the independent variables, are all equal. For internal estimation, an additional assumption is that the conditional distributions for the dependent variable follow the normal probability distribution.

Deviations Taken from Actual Means

The work involved in finding these regression equations is reduced by taking deviations from the means of the variables under consideration. The regression equation for three variables then becomes:

$$x_1 = b_{12.3} + b_{13.2}\, x_3$$

where $x_1 = (X_1 - \overline{X}_1)$, $x_2 = (X_2 - \overline{X}_2)$, $x_3 = (X_3 - \overline{X}_3)$

The value of $b_{12.3}$ and $b_{13.2}$ can be obtained by solving simultaneously the following two normal equations:

$$\sum x_1 x_2 = b_{12.3} \sum x_2^2 + b_{13.2} \sum x_2 x_3$$

$$\sum x_1 x_2 = b_{12.3} \sum x_2 x_3 + b_{13.2} \sum x_3$$

The value of $b_{12.3}$ and $b_{13.2}$ can also be obtained as follows:

$$b_{12.3} = r_{12.3} \times \frac{\sigma_{1.23}}{\sigma_{2.13}}$$

$$b13.2 = r_{13.2} \times \frac{\sigma_{13.2}}{\sigma_{3.12}}$$

The regression equation of X_1 on X_2 and X_3 can be expressed as follows:

$$(x_1 - \overline{X}_1) = \left(\frac{r_{12} - r_{13}\, r_{23}}{1 - r^2_{23}}\right)\left(\frac{S_1}{S_2}\right)(x_2 - \overline{X}_2) + \left(\frac{r_{13} - r_{12}\, r_{23}}{1 - r^2_{23}}\right)\left(\frac{S_1}{S_3}\right)(x_3 - \overline{X}_3)$$

The regression equation of X_3 on X_2 and X_1 can be written as follows:

$$(x_3 - \overline{X}_3) = \left(\frac{r_{23} - r_{13}\, r_{12}}{1 - r^2_{12}}\right)\left(\frac{S_3}{S_2}\right)(x_2 - \overline{X}_2) + \left(\frac{r_{13} - r_{23}\, r_{12}}{1 - r^2_{12}}\right)\left(\frac{S_3}{S_1}\right)(x_1 - \overline{x}_1)$$

This method of obtaining regression equation is much simpler as compared to one where simultaneously several normal equations are to be solved. For calculating regression equation for three variables when the above procedure is used, we need the following:

$\overline{X}_1$ $\quad \overline{X}_2$ $\quad \overline{X}_3$

S_1 $\quad S_2$ $\quad S_3$

r_{12} $\quad r_{13}$ $\quad r_{..}$

Other Equations of Multiple Linear Regression

In the case of two variables, there were two equations of regression—one of them indicating regressions of Y on X and the other, that of X on Y. When there are three variables, there will be three equations of regression, one indicating the regression of X_1 on X_2 and X_3, the other indicating the regression of X_2 on X_1 and X_3 and third indicating the regression of X_3 on X_1 and X_2. The first of these has been given earlier. If X_2 and X_3 were to be treated as dependent variables, the regression equation will respectively be:

$$X_2 = a_{2.13} + b_{21.3}\ X_1 + b_{23.1}\ X_3 \qquad \text{...(ii)}$$

$$X_3 = a_{3.12} + b_{31.2}\ X_1 + b_{32.1}\ X_2 \qquad \text{...(iii)}$$

The normal equations for fitting equation (ii) will be:

$$\sum X_2 = N a_{2.13} + b_{21.3} \sum X_1 + b_{23.1} \sum X_3$$

$$\sum X_1X_2 = a_{2.13}\sum X_1 + b_{21.3} \sum X_1^2 + b_{23.1} \sum X_1X_2$$

$$\sum X_2X_3 = a_{2.13}\sum X_3 + b_{21.3} \sum X_1X_3 + b_{23.1} \sum X_3^2$$

In case we want to fit equation (iii) the normal equations will be:

$$\sum X_3 = N a_{3.12} + b_{31.2} \sum X_1 + b_{32.1} \sum X_2$$

$$\sum X_1X_3 = a_{3.12}\sum X_2 + b_{31.2} \sum X_1^2 + b_{32.1} \sum X_1X_2$$

$$\sum X_2X_3 = a_{3.12}\sum X_3 + b_{31.2} \sum X_1X_2 + b_{32.1} \sum X_2^2$$

Generalizations for More Than Three Variables

In case of four variables, the linear regression equation of X_1 on X_2, X_3 and X_4 can be written as

$$X_1 = a_{1.234} + b_{12.34}\ X_2 + b_{13.24}\ X_3 + b_{14.23}\ X_4$$

It represents a hyperplane in four dimensional space. On formal multiplication of both sides of the above equation by X_1, X_2, X_3 and X_4 successively and then summing on both sides, we obtain the normal equations for determination of $a_{1.234}$, $b_{13.24}$, $b_{12.34}$ and $b_{14.23}$ which when substituted to above equation, gives the least square regression equation of X_1 on X_2, X_3 and X_4.

While drawing statistical inference in multiple regression, it should be noted that the regression coefficients for highly intercorrelated independent variables tend to be unreliable. This is for the reason that when independent variables are high intercorrelated, it is extremely difficult to separate out the individual influences of each variable. There is a great deal of concern in fields such as econometrics and applied statistics with this problem of intercorrelation among dependent variables, often referred to as multi-collinearity. As suggested by Morris Hamburg, one of the simplest solutions to the problem of two highly correlated independent variables is merely to discard one of the variables.

Use of Computers in Multiple Regression and Correlation Analysis

The application of multiple regression and correlation analysis requires extensive and highly precise computations. As the number of variables increases, the computations become more and more difficult and time consuming. The computers are being used widely in the application of these techniques. A large number of computer installations have one or more multiple regression and correlation programmes in the programme library that are available to users. In fact, it is becoming increasingly unnecessary nowadays to carry out regression analysis by hand.

The availability of these programmes enables many analysts to obtain the desired regression and correlation result without the analyst having to spend time writing a computer programme. The suitability of a given library programme for use in a particular problem depends upon the input requirements, operating procedure and results computed by the programme. Many library programmes are sufficiently general and comprehensive to fulfil the requirements of a wide variety of users.

It may be pointed out that though with the use of computers it is possible to test and include large number of independent variables in a regression analysis, good judgment and knowledge of the logical relationships involved must always be used as a guide to deciding which variables to include in the construction of a regression equation.

Example:

Given the following, determine the regression equation of

(i) X_1 *on* X_2 *and* X_3

(ii) X_2 *on* X_1 *and* X_3

$r_{12} = 0.8 r_{13} = 0.6; r_{23} = 0.5$

$\sigma_1 = 10; \sigma_2 = 8; \quad \sigma_3 = 5.$

Solution:

(i) Regression equation of X_1 on X_2 and X_3 is given by

$$X_1 = a_{1.23} + b_{12.3} X_2 + b_{13.2} X_3$$

It the variates X_1, X_2 and X_3 are measured as deviations from their respective means, 'a' will be zero. The values of $b_{12.3}$ and $b_{13.2}$ can be calculate from the data given above but not for 'a'. So let us assume X_1, X_2 and X_3 represent deviations from means. So the regression equation of X_1 on X_2 and X_3 is

$$X_1 = b_{12.3} X_2 + b_{13.2} X_3$$

$$b_{12.3} = \frac{\sigma_1}{\sigma_2} \times \frac{r_{12} - r_{12} r_{23}}{1 - r^2_{23}}$$

$$= \frac{10}{8} \times \frac{0.8 - (0.6)(0.5)}{1 - (0.5)^2} = 0.833$$

$$b_{13.2} = \frac{\sigma_1}{\sigma_2} \times \frac{r_{13} - r_{12} r_{23}}{1 - r^2_{23}}$$

$$= \frac{10}{5} \times \frac{0.6 - (0.8)(0.5)}{1 - (0.5)^2} = 0.533$$

$\therefore$ Required regression equation is

$$X_1 = 0.833 X_2 + 0.533 X_3$$

(ii) Regression equation of X_2 on X_1 and X_3

$$X_2 = b_{12.3} X_1 + b_{23.1} X_3$$

$$b_{12.3} = \frac{\sigma_2}{\sigma_1} \times \frac{r_{12} - r_{23} r_{13}}{1 - r^2_{13}}$$

$$= \frac{8}{10} \times \frac{0.8 - (0.5)(0.6)}{1 - (0.6)^2}$$

$$= \frac{8}{10} \times \frac{0.8 - 0.3}{1 - .36}$$

$$= \frac{8}{10} \times \frac{0.5}{0.64} = 0.625$$

$$b_{23.1} = \frac{\sigma_2}{\sigma_3} \times \frac{r_{23} - r_{12} r_{13}}{r^2_{13}}$$

$$= \frac{8}{5} \times \frac{0.5 - (0.8)(0.6)}{1 - (0.6)^4}$$

$$= \frac{8}{5} \times \frac{0.02}{0.64} = 0.05$$

Thus, $X_2 = 0.625\ X_1 + 0.05\ X_3$.

RELIABILITY OF ESTIMATES

The problem of determining the accuracy of estimates from the multiple regression is basically the same as for estimates from a simple regression equation. Since the correlation is seldom perfect, estimates made from the regression equation will deviate from the correct value of the dependent variable. If an estimate is to be fo maximum usefulness, it is necessary to have some indication of its precision. Just as with the simple regression equation, the measure of reliability is an average of these deviations of the actual value of non-dependent variable from the estimate from the regression equation or, in other words, the standard error estimate.

The standard error of estimate of X_1 on X_2 and X_3 is defined as

$$S_{1.23} = \sqrt{\frac{\sum(X_1 - X_{last})^2}{N-3}}$$

$S_{1.23}$ represents standard error estimate of X_1 on X_2 and X_3. X_{last} indicates the estimated value of X_1 as calculated from the regression equations.

In terms of the correlation coefficients r_{12}, r_{13} and r_{23}, the standard error of estimate can also be computed from the result:

$$S_{1.23} = S_1\sqrt{\frac{1 - r^2{}_{12} - r^2{}_{13} - r^2{}_{23} + 2\,r_{23}\,r_{13}\,r_{23}}{1 - r^2{}_{23}}}$$

The standard error measures the closeness of estimates derived from the regression equation to actual observed values.

Coefficient of Multiple Determination

The coefficient of multiple determination is analogous to the coefficient of determination in the two-variable case. As explained earlier, the fit of a straight line to the two-variable scatter was measured by the simple coefficient of determination r^2 which was defined as the ratio of the explained sum of squares to the total sum of squares. In the same fashion, we can define coefficient of multiple determination which is denoted by R^2. Symbolically:

$$R^2 = \frac{SSR}{SST} = 1 - \frac{SSE}{SST}$$

Similar to the case of r^2 and r in two variable analysis, R^2 is easier to interpret since R^2 is a percentage figure whereas R is not.

As a ratio of explained variation to the total variation in X^1, R^2 can be interpreted as the proportion of the total variation in the dependent variable that is associated with or explained by the regression of X_1 on X_2 and X_3. We may also think of R^2 as a measure of closeness of fit of the regression plane to the actual points. The closer the value of R^2 to 1, the smaller is the scatter of the points about the regression plane and the better is the fit.

The square root of the coefficient of multiple determination is called the coefficient of multiple correlation, denoted as R. This measure is seldom used in practice.

Example 1:

Find the multiple linear regression equation of X_1 on X_2 and X_3 from the data relating to three variables given below:

X_1	4	6	7	9	13	15
X_2	15	12	8	6	4	3
X_3	30	24	20	14	10	4.

Solution:

The regression equation of X_1 on X_2 and X_3 is

$$X_1 = a_{1.23} + b_{12.3} X_2 + b_{13.2} X_3.$$

The value of the constants $a_{1.23}$, $b_{12.3}$ and $b_{13.2}$ are obtained by solving the following three normal equations:

$$\Sigma X_1 = Na_{1.23} + b_{12.3} \Sigma X_2 + b_{13.2} \Sigma X_3$$

$$\Sigma X_1 X_2 = a_{1.23} \Sigma X_2 + b_{12.3} \Sigma X_2^2 + b_{13.2} \Sigma X_2 X_3$$

$$\Sigma X_1 X_3 = a_{1.23} \Sigma X_3 + b_{12.3} \Sigma X_2 X_3 + b_{13.2} \Sigma X_3^2$$

Calculating the required values:

X_1	X_2	X_3	X_1X_2	X_1X_3	X_2X_3	X_2^2	X_3^2	X_1^2
4	15	30	60	120	450	225	900	16
6	12	24	72	144	288	144	576	36
7	8	20	56	140	160	64	400	49
9	6	14	54	126	84	36	196	81
13	4	10	52	130	40	16	100	169
15	3	4	45	60	12	9	16	225
ΣX_1 =54	ΣX_2 =48	ΣX_3 =102	ΣX_1X_2 =339	ΣX_1X_3 =720	ΣX_2X_3, =1,034	ΣX_2^2 =494	ΣX_3^2 =2,188	ΣX_1^2 =576

Substituting the values in the normal equations:

$$6\ a_{1.23} + 48\ b_{12.3} + 102\ b_{13.2} = 54 \quad \text{...(i)}$$

$$48\ a_{1.23} + 494\ b_{12.3} + 103\ b_{13.2} = 339 \quad \text{...(ii)}$$

$$102\ a_{1.23} + 1034\ b_{12.3} + 2188\ b_{13.2} = 720 \quad \text{...(iii)}$$

Multiplying Eq. (i) by 8, we get

$$48a_{1.23} + 384\ b_{12.3} + 816\ b_{13.2} = 432 \quad \text{...(iv)}$$

Subtracting Eqn. (ii) from (iv), we get

$$110\ b_{12.3} + 218\ b_{13.2} = 93 \quad \text{...(v)}$$

Multiplying Eqn. (i) by 17, we get

$$102a_{1.23} + 816\ b_{12.3} + 1734\ b_{13.2} = 918 \quad \text{...(vi)}$$

Subtracting Eqn. (iii) from Eqn. (vi), we get

$$218\ b_{12.3} + 454\ b_{13.2} = -198 \quad \text{...(vii)}$$

Multiplying Eq. (v) by 109, we obtain

$$11990\ b_{12.3} + 23762\ b_{13.2} = 10137 \quad \text{...(viii)}$$

Multiplying Eqn. (vii) by 55, we get

$$11990\ b_{12.3} + 23762\ b_{13.2} = -10890 \quad \text{...(ix)}$$

Subtracting Eqn. (viii) from Eqn. (ix), we get

$$1208\ b_{13.2} = -753$$

$$\text{or } b_{13.2} = \frac{-753}{1208} = -0.623$$

Substituting the value of $b_{13.2}$ in Eqn. (v), we get

$$110\ b_{12.3} = 218(-0.623) = -93$$

$$110\ b_{12.3} = 135.814 - 93$$

$$b_{12.3} = \frac{42814}{110} = +0.389$$

Substituting the values of $b_{13.2}$ in Eqn. (i), we get

$$6a_{1.23} + 48\ (0.389) + 102\ (-0.623) = 54$$

$$6a_{1.23} = 54 + 63.564 - 18.672 = 98.874$$

$$a_{1.23} = 16.479$$

Thus, the required regression equation is:

$$X_1 = 16.479 + 0.389\ X_2 - 0.623\ X_3$$

Example 2:

In a trivariate distribution:

$\sigma_1 = 3,\ \sigma_2 = 4,\ \sigma_3 = 5$

$r_{23} = 0.4,\ r_{31} = 0.6,\ r_{13} = 0.7$

Determine the regression equation of X_1 on X_2 and X_3 if the variates are measured from their means.

Solution:

When the variates are measured from means, the regression equation of X_1, X_2 and X_3 is given by:

$$X_1 = b_{12.3} X_2 + b_{13.2} X_3$$

$$b_{12.3} = \frac{\sigma_1}{\sigma_2} \times \frac{r_{12} - r_{13} r_{23}}{1 - r^2_{23}}$$

Substituting the given values

$$= \frac{3}{4} \times \frac{.7 \times .6 \times .4}{1 - (.4)^2}$$

$$= \frac{3}{4} \times \frac{.46}{.84} - 0.411$$

$$b_{13.2} = \frac{\sigma_1}{\sigma_3} \times \frac{r_{13} - r_1 r_{23}}{1 - r^2_{23}}$$

$$= \frac{3}{5} \times \frac{.6 - .7 \times .4}{1 - (.4)^2}$$

$$= \frac{3}{5} \times \frac{.32}{.84} = 0.229$$

Hence the required equation is

$$X_1 = 0.41\ X_2 + 0.229\ X_3$$

Example 3:

The simple correlation coefficients between temperature (X_1), com yield (X_2) and rainfall (X_3) are $r_{12} = 0.59$, $r_{13} = 0.46$ and $r_{23} = 0.77$. Calculate partial correlation coefficient $r_{12.3}$ and multiple correlation coefficient $R_{1.23}$.

Solution:

$$r_{12.3} = \frac{r_{12} - r_{13} r_{23}}{\sqrt{1 - r^2_{13}}\sqrt{1 - r^2_{23}}}$$

$$r_{12} = .59,\ r_{13} = .46,\ r_{23} = .77$$

$$r_{12.3} = \frac{.59 - .46 \times .77}{\sqrt{1-(46)^2}\sqrt{1-(.77)^2}}$$

$$= \frac{.59 - .3542}{\sqrt{1 - .2116}\sqrt{1 - .5529}} = \frac{.2358}{.5665} = 0.416$$

$$R_{1.32} = \frac{\sqrt{r^2{}_{12} + r^2{}_{13} - 2\, r_{12}\, r_{13}\, r_{23}}}{1 - r^2{}_{23}}$$

$$= \frac{\sqrt{(.59)^2 + (.46)^2 - 2(.59 \times .46 \times .77)}}{1-(.77)^2}$$

$$= \frac{\sqrt{.3481 + .2116 - .418}}{.4071}$$

$$= \frac{\sqrt{.5597 - .418}}{.4071} = \frac{\sqrt{.1417}}{.4071} = .589$$

(b) *From the following information, calculate $R_{1.23}$*
$r_{12} = 0.8,\ r_{13} = 0.5,\ r_{23} = 0.3$

Solution:

$$R_{1.23} = \frac{\sqrt{r^2{}_{12} + r^2{}_{13} - 2\, r_{12}\, r_{13}\, r_{23}}}{1 - r^2{}_{23}}$$

$$= \frac{\sqrt{(.8)^2 + (.5)^2 - 2(.8)(.5)(.3)}}{1-(.3)^2}$$

$$= \frac{\sqrt{.64 + .25 - .24}}{.91} = 0.845$$

Example 4:

Explain multiple and partial correlation in a trivariate distribution $r_{12} = 0.863$, $r_{12} = 0.648$ and $r_{23} = 0.709$.

Find $r_{12.3}$ and $R_{13.2}$.

Solution:

$$r_{12.3} = \frac{\sqrt{r_{12} - r_{13}\, r_{23}}}{\sqrt{1 - r^2{}_{13}}\sqrt{1 - r^2{}_{23}}}$$

$$r_{12} = .863,\ r_{13} = .648,\ r_{23} = .709$$

$$r_{12.3} = \frac{.863 - .648 \times .709}{\sqrt{1-(.648)^2}\sqrt{1-(.709)^2}}$$

$$= \frac{.863 - .4594}{\sqrt{.580 \times .497}} = \frac{.4036}{\sqrt{.2883}} = \frac{.4036}{.537} = 0.752$$

$$R_{1.23} = \frac{\sqrt{r^2{}_{12} + r^2{}_{13} - 2r_{12}\, r_{13}\, r_{23}}}{1 - r^2{}_{23}}$$

$$= \frac{\sqrt{(.863)^2 + (.648)^2 - 2(.863)(.648)(.789)}}{1-(.709)}$$

$$= \frac{\sqrt{.745 + .42 - .793}}{.467} = 0.687$$

Example 5:

If $r_{12} = 0.8$, $r_{13} = 0.4$ and $r_{23} = 0.56$, find the value of $r_{23.2}$, $r_{13.2}$ and $r_{23.1}$.

Solution:

$$r_{12.3} = \frac{r_{12} - r_{13}\, r_{23}}{\sqrt{1 - r^2{}_{23}}\sqrt{1 - r^2{}_{23}}}$$

Substituting te given values

$$r_{12.3} = \frac{.8 - .4 \times .56}{\sqrt{1-(.4)^2}\sqrt{1-(.56)^2}}$$

$$= \frac{.8 - .224}{\sqrt{1 - .16}\sqrt{1 - .316}} = \frac{.576}{\sqrt{.84 \times .6864}}$$

$$= \frac{.576}{.759} = 0.759$$

$$r_{13.2} = \frac{r_{13} - r_{12}\, r_{23}}{\sqrt{1 - r^2{}_{12}}\sqrt{1 - r^2{}_{23}}}$$

$$= \frac{.4 - .8 \times .56}{\sqrt{1-(.8)^2}\sqrt{1-(.56)^2}}$$

$$= \frac{.4-.448}{\sqrt{1-.64}\sqrt{1-3136}} = \frac{-.048}{0.6 \times 0.828}$$

$$= \frac{-.048}{.497} = -0.097$$

$$r_{23.1} = \frac{r_{23} - r_{12}\, r_{13}}{\sqrt{1-r^2{}_{12}}\sqrt{1-r^2{}_{13}}}$$

$$= \frac{.56-.8 \times .4}{\sqrt{1-(.8)^2}\sqrt{1-(.4)^2}}$$

$$= \frac{.56-.32}{\sqrt{36 \times .84}} = \frac{.24}{.55} = 0.436.$$

Example 6:

In a trivariate distribution

$\sigma_1 = 2,\ \sigma_2 = \sigma_3 = 3$

$r_{12} = 0.7,\ r_{23} = r_{31} = 0.5$

Find (i) $b_{12.3}$ *and (ii)* $b_{13.2.}$

Solution:

(i) $$b_{12.3} = r_{12.3}\frac{\sigma_{1.3}}{\sigma_{2.3}}$$

$$r_{12.3} = \frac{r_{12} - r_{13}\, r_{23}}{\sqrt{1-r^2{}_{13}}\sqrt{1-r^2{}_{23}}} = \frac{0.7-(0.5)(0.5)}{\sqrt{1-(0.5)^2}\sqrt{1-(0.5)^2}}$$

$$= \frac{0.7-0.25}{\sqrt{0.75}\sqrt{0.75}} = \frac{0.45}{0.75} = 0.6$$

$$\sigma_{1.3} = \sigma_1\sqrt{\left(1-r^2{}_{23}\right)} = 2\sqrt{1-0.25} = 1.732$$

$$s_{2.3} = \sigma_2\sqrt{\left(1-r^2{}_{23}\right)} = 3\sqrt{1-0.25} = 2.598$$

$$b_{12.3} = 0.6 \times \frac{1.732}{2.598} = 0.4$$

$$b_{13.2} = r_{13.2}\frac{\sigma_{1.2}}{\sigma_{3.2}}$$

$$r_{13.2} = \frac{r_{13} - r_{12\ 23}}{\sqrt{1-r^2_{12}}\sqrt{1-r^2_{23}}} = \frac{0.5-(0.7\times 0.5)}{\sqrt{1-(0.7)^2}\sqrt{1-(0.5)^2}}$$

$$= \frac{0.15}{\sqrt{0.51}\sqrt{0.75}} = 0.243$$

$$\sigma_{1.2} = \sigma_1 \sqrt{(1-r^2_{13})} = 2\sqrt{1-0.49} = 1.428$$

$$\sigma_{3.2} = \sigma_3 \sqrt{(1-r^2_{23})} = 3\sqrt{1-0.25} = 2.598$$

$$b^{13.2} = 0.243 \times \frac{1.428}{2.598} = 0.134$$

Example 7:

Given $r_{12} = 0.28$, $r_{23} = 0.49$, $r_{31} = 0.51$

$$\sigma_1 = 2.7,\ \sigma_2 = 2.4,\ \sigma_3 = 2.7$$

Find the regression equation of X_3 on X_1 and X_2.

Solution:

The regression equation of X_3 on X_1 and X_2 is

$$X_3 = b_{31.2}\, X_1 + b_{32.1}\, X_2$$

$$b_{31.2} = \frac{r_3}{\sigma_1} \times \frac{r_{31} - r_{12}r_{23}}{1-r^2_{12}}$$

$$= \frac{2.7}{2.7} \times \frac{.51-(.28)(.49)}{1-(.28)^2} = 0.405$$

$$b_{32.1} = \frac{\sigma_3}{\sigma_2} \times \frac{r_{23} - r_{12}\, r_{13}}{1-r^2_{12}}$$

$$= \frac{2.7}{2.4} \times \frac{.49-(2.8)(.51)}{1-(.28)^2} = 0.424$$

$$X_3 = 0.405\, X_1 + 0.424\, X_2$$

Example 8:

In a trivariate distribution,

$\sigma_1 = 3$, $\sigma_2 = \sigma_3 = 5$

$r_{12} = 0.6,\ r_{23} = r_{31} = 0.8$

Find (i) $r_{23.1}$ *and (ii)* $R_{1.23}$.

Solution:

(i) $$r_{23.1} = \frac{r_{23} - r_{12}\, r_{13}}{\sqrt{1-r^2_{12}}\,\sqrt{1-r^2_{13}}}$$

$$= \frac{0.8 - 0.6 \times 0.8}{\sqrt{1-(0.6)^2}\,\sqrt{1-0.8^2}} = \frac{0.8 - 0.48}{\sqrt{0.64}\,\sqrt{0.36}} = 0.667$$

(ii) $$R_{1.23} = \frac{\sqrt{r^2_{12} + r^2_{13} - 2r_{12}\, r_{13}\, r_{23}}}{1-r^2_{23}}$$

$$= \frac{\sqrt{(0.6)^2 + (0.8)^2 - 2(0.6)(0.8)(0.8)}}{1-(0.8)^2}$$

$$= \frac{\sqrt{0.36 + 0.64 - 0.768}}{0.36} = \sqrt{0.644} = 0.803.$$

Example 9:

Calculate (a) $R_{1.23}$, *(b)* $R_{3.12}$ *and (c)* $R_{2.13}$ *for the following data:*

$\overline{X}_1 = 6.8 \qquad \overline{X}_2 = 7.0 \qquad \overline{X}_3 = 74$

$S_1 = 1.0 \qquad S_2 = 0.8 \qquad S_3 = 9$

$r_{12} = 0.6 \qquad r_{13} = 0.7 \qquad r_{23} = 0.65$

Solution:

$$R_{1.23} = \frac{r^2_{12} + r^2_{13} - 2r_{12}r_{13}r_{23}}{1-r^2_{12}}$$

$$= \frac{\sqrt{(0.6)^2 + (0.7)^2 - (2 \times 0.6 \times 0.7 \times 0.65)}}{1-(0.65)^2}$$

$$= \frac{\sqrt{0.36 + 0.49 - 0.546}}{0.5775} = \sqrt{0.526} = 0.725$$

$$R_{3.12} = \frac{\sqrt{r^2_{12} + r^2_{23} - 2r_{12}r_{13}\, r_{23}}}{1-r^2_{12}}$$

$$= \frac{\sqrt{(0.7)^2 + (0.65)^2 - (2 \times 0.6 \times 0.7 \times 0.65)}}{1-(0.6)^2}$$

$$= \frac{\sqrt{0.49 + 0.4225 - 0.546}}{1-0.36} = \sqrt{0.573} = 0.757$$

$$R_{2.13} = \frac{\sqrt{r^2_{12} + r^2_{23} - 2r_{12}\, r_{13}\, r_{23}}}{1-r^2_{13}}$$

$$= \frac{\sqrt{(0.6)^2 + (0.65)^2 - (2 \times 0.6 \times 0.7 \times 0.65)}}{1-(0.7)^2}$$

$$= \frac{\sqrt{0.36 + 0.4225 - 0.546}}{0.51} = \sqrt{0.464} = 0.681.$$

Example 10:

The correlation between a general intelligence test and school achievement in a group of children from 6 to 15 years old is 0.80. The correlation between the general intelligence test and age in the same group is 0.70 and the correlation between school achievement and age is 0.60. What is the correlation between general intelligence and school achievement in children of the same age? Comment on the result.

Solution:

Let X_1 denote general intelligence test

X_2 denote school achievement

X_3 denote age.

We are given:

$r_{12} = 0.8$, $r_{13} = 0.7$ and $r_{23} = 0.6$

We are find $r_{12.3}$

$$r_{12.3} = \frac{r_{12} - r_{13}\, r_{23}}{\sqrt{1-r^2_{13}}\sqrt{1-r^2_{23}}}$$

$$= \frac{0.8 - 0.7 \times 0.6}{\sqrt{1-(0.7)^2}\sqrt{1-(0.6)^2}}$$

$$= \frac{0.8 - 0.42}{\sqrt{0.51}\sqrt{0.64}} = \frac{0.38}{0.57} = 0.667$$

Example 11:

An instructor of mathematics wishes to determine the relationship of grades on a final examination of grades on two quizes given during the semester. Costing X_1, X_2 and X_3 the grades of a student on the first quiz and final examination respectively, he made the following computations for a total of 120 students:

$$\overline{X}_1 = 6.8 \qquad \overline{X}_2 = 0.7 \qquad \overline{X}_3 = 74$$

$$S_1 = 1.0 \qquad S_2 = 0.80 \qquad S_3 = 9.0$$

$$r_{12} = 0.60 \qquad r_{13} = 0.70 \qquad r_{23} = 0.65$$

(i) Find the least square regression equation of X_3 on X_1 and X_2.

(ii) Estimate the final grades of two students who scored respectively 9 and 7, 4 and 5 on the two quizes.

Solution:

The regression equation of X3 and X2 on X1 can be written as:

$$\left(X_3 - \overline{X}_3\right) = \left(\frac{r_{23} - r_{12}\, r_{13}}{1 - r^2_{12}}\right)\left(\frac{S_3}{S_2}\right)\left(X_2 - \overline{X}_2\right)$$

$$+ \left(\frac{r_{13} - r_{23}\, r_{12}}{1 - r^2_{12}}\right)\left(\frac{S_3}{S_1}\right)\left(X_1 - \overline{X}_1\right)$$

Substituting the given values:

$$(X_3 - 74) = \left(\frac{0.65 - (0.7 \times 0.6)}{1 - (0.6)^2}\right)\left(\frac{9}{0.8}\right)(X_2 - 7) + \left(\frac{0.7 - (0.65 \times 0.6)}{1 - (0.6)^2}\right)$$

$$\left(\frac{9}{1}\right)(X_1 - 6.8)$$

$$(X_3 - 74) = \left(\frac{0.65 - 0.42}{0.64}\right)\left(\frac{9}{0.8}\right)(X_2 - 7) + \left(\frac{0.7 - 0.39}{0.64}\right)(9)(X_1 - 6.8)$$

$$(X_3 - 74) = 4.04\ (X_2 - 7) + 4.36\ (X_1 - 6.8)$$

$$(X_3 - 74) = 4.04\ X_2 - 28.28 + 4.36\ X_1 - 29.65$$

$$X_3 = 16.07 + 4.36\ X_1 + 4.04\ X_2$$

Final grades of students who scored 9 and 7 marks:

$X_1 = 9, X_2 = 7$

$X_3 = 16.07 + 4.36\ (9) + 4.04\ (7)$

$= 16.04 + 17.44 + 32.32 = 65.8.$

Example 12:

If $r_{12} = 0.9,\ r_{13} = 0.75,\ r_{23} = 0.7$, *find* $R_{1.23}$.

Solution: (a)

$$R1.23 = \frac{\sqrt{r_{12}^2 + r_{13}^2 - 2r_{12}\,r_{13}\,r_{23}}}{1 - r_{23}^2}$$

Substituting the given values

$$r_{1.23} = \frac{\sqrt{(0.9)^2 + (0.75)^2 - 2 \times 0.9 \times 0.75 \times 0.7}}{1-(0.7)^2}$$

$$= \frac{\sqrt{0.81 + 0.5625 - 0.945}}{0.51} = \frac{\sqrt{0.4275}}{0.51} = \sqrt{0.838} = 0.916$$

(b) *Given that* $r_{12} = 0.77,\ r_{13} = 0.72,\ r_{23} = 0.52$; *calculate* $R_{1.23}$.

$$R_{1.23} = \frac{\sqrt{r_{12}^2 + r_{13}^2 - 2\,r_{12}\,r_{13}\,r_{23}}}{1 - r_{23}^2}$$

$r_{12} = 0.77,\ r_{13} = 0.72,\ r_{23} = 0.52$

Substituting the values

$$R_{1.23} = \frac{\sqrt{(.77)^2 + (.72)^2 - 2(.77)(.72)(.52)}}{1-(.52)^3}$$

$$= \frac{\sqrt{.5929 + .5184 - .577}}{1 - .2704} = \frac{\sqrt{.5343}}{.7296} = 0.856$$

Example 13:

Given the following information:

$r_{12} = 0.20,\ r_{13} = 0.40,\ r_{23} = 0.50$

$r_{14} = 0.40,\ r_{24} = 0.30,\ r_{34} = 0.1$

Find $r_{41.23}$.

Solution:

$$r_{41.23} = \frac{r_{41.3} - r_{12.3}\, r_{24.3}}{\sqrt{1-r^2_{12.3}}\sqrt{1-r^2_{24.3}}}$$

We have to find first $r_{41.3}$, $r_{12.3}$ and $r_{24.3}$

$$r_{41.3} = \frac{r_{14} - r_{13}\, r_{34}}{\sqrt{1-r^2_{13}}\sqrt{1-r^2_{34}}} = \frac{0.4-(0.4\times -0.1)}{\sqrt{1-(0.4)^2}\sqrt{1-(0.1)^2}}$$

$$= \frac{0.4-0.04}{\sqrt{0.84}\sqrt{0.99}} = 0.482$$

$$r_{12.3} = \frac{r_{12} - r_{13}\, r_{23}}{\sqrt{1-r^2_{13}}\sqrt{1-r^2_{23}}} = \frac{0.02-(0.4\times 0.5)}{\sqrt{1-(0.4)^3}\sqrt{1-(0.5)^2}}$$

$$= \frac{-0.2-0.2}{\sqrt{0.84}\sqrt{0.75}} = 0.504$$

$$r_{24.3} = \frac{r_{24} - r_{34}\, r_{23}}{\sqrt{1-r^2_{24}}\sqrt{1-r^2_{23}}} = \frac{0.3-(-0.1\times 0.5)}{\sqrt{1-(-0.1)^2}\sqrt{1-(0.5)^2}}$$

$$= \frac{0.3+0.05}{\sqrt{0.99}\sqrt{0.75}} = \frac{0.3}{0.862} = 0.406$$

Substituting the values:

$$r_{41.23} = \frac{0.482-(-0.504\times 0.406)}{\sqrt{1-(0.504)^2}\sqrt{1-(0.406)^2}} = 0.871$$

Example 14:

Given $r_{12} = 0.5$, $r_{13} = 0.4$ and $r_{23} = 0.1$, find $r_{12.3}$ and $r_{23.1}$.

Solution:

$$\frac{r_{12} - r_{13}\, r_{23}}{\sqrt{1-r^2_{13}}\sqrt{1-r^2_{23}}}$$

Substituting the values:

$$r_{12.3} = \frac{.5-.4\times .1}{\sqrt{1-(.4)^2}\sqrt{1-(.1)^2}}$$

$$= \frac{.5-.04}{\sqrt{.84}\sqrt{.99}} = \frac{.46}{.912} = 0.504$$

$$r_{23.1} = \frac{r_{23} - r_{12}\, r_{13}}{\sqrt{1-r^2_{12}}\sqrt{1-r^2_{13}}}$$

$$= \frac{.1 - .5 \times .4}{\sqrt{1-(.5)^2}\sqrt{1-(.4)^2}}$$

$$= \frac{-.1}{\sqrt{.75}\sqrt{.84}} = \frac{-.1}{.794} = -\ 0.126.$$

Example 15:

The following constants are obtained from measurements on length in mm. (X_1), volume in c.c. (X_2) and weight in gm. (X_2) of 300 eggs:

$\overline{X}_1 = 55.95 \qquad S_1 = 2.26 \quad r_{12} = 0.578$

$\overline{X}_2 = 51.48 \qquad S_2 = 4.39 \quad r_{13} = 0.581$

$\overline{X}_3 = 56.03 \qquad S_3 = 4.41 \quad r_{23} = 0.974$

Obtain the linear regression equation of egg weight on egg length and egg volume. Hence estimate the weight of an egg whose length is 58 mm. and volume is 52.5 c.c.

Solution:

We have to obtain linear regression equation of egg weight on egg length and egg volume, *i.e.*, X_3 on X_1 and X_2. The regression equation of X_2 on X_3 and X_1 can be written as:

$$X_3 - \overline{X}_3 = \left(\frac{r_{23} - r_{13}\, r_{12}}{1-r^2_{12}}\right)\left(\frac{S_3}{S_2}\right) X_2 - \overline{X}_2 + \left(\frac{r_{13} - r_{23}\, r_{12}}{1-r^2_{13}}\right)\left(\frac{S_3}{S_1}\right)(X_1 - \overline{X}_1)$$

Substituting the values

$$X_3 - 56.03 = \left(\frac{0.974 - (0.581 \times 0.578)}{1-(0.578)^2}\right)\left(\frac{4.41}{4.39}\right)$$

$$(X_2 - 51.48) + \left[\frac{0581 - (0.974 \times 0.578)}{1-(0.578)^2}\right]\left(\frac{4.41}{2.26}\right) \quad (X_2 - 55.95)$$

$$X_3 - 56.03 = \left(\frac{0.974 - 0.335}{1 - 0.334}\right)\left(\frac{4.41}{4.39}\right) \quad (X_2 - 51.48)$$

$$+ \left(\frac{0.581 - 0.563}{1 - 0.334}\right)\left(\frac{4.41}{2.26}\right) \quad (X_1 - 55.95)$$

$$X_3 - 56.03 = 0.964\ (X_2 - 51.48)\ 0.053\ (X_1 - 55.95)$$

$$X_3 - 56.03 = 0.964\ X_2 - 49.63 + 0.053\ X_1 - 2.97$$

$$X_3 = 3.43 + 0.053\ X_1 + 0.964\ X_2$$

When length, *i.e.*, X_1 is 58 and volume, *i.e.*, X_2 is 52.5, the weight of the egg would be:

$$X_3 = 3.43 + 0.053\ (58) + 0.964\ (52.5)$$

$$= 3.43 + 3.074 + 50.61 = 57.114 \text{ or } 57.11 \text{ gms.}$$

Example 16:

Suppose, a computer has found, for a given set of values of X_1, X_2 and X_3,

$r_{12} = 0.91$, $r_{13} = 0.33$ *and* $r_{23} = 0.81$

explain whether these computations may be said to be free from errors.

Solution:

For determining whether the given computations are correct or not, we calculate the value of $r_{12.3}$. If the value of $r_{12.3}$ is less than one, the computations can be regarded as free errors.

$$r_{12.3} = \frac{r_{12} - r_{12}\, r_{23}}{\sqrt{1 - r^2_{13}}\sqrt{1 - r^2_{23}}}$$

$$= \frac{.91 - (.33)(.81)}{\sqrt{1 - (.33)^2}\sqrt{1 - (.81)^2}}$$

$$= \frac{.91 - .2673}{\sqrt{1 - .1089}\sqrt{1 - .6561}} = \frac{.6427}{\sqrt{.8911 \times .3439}}$$

$$= \frac{.6427}{.5536} = 1.161$$

Since the value of $r_{12.3}$ cannot exceed one, the computations given in the question are not free from errors.

Example 17:

If $r_{13} = 0.65$, $r_{23} = 0.6$ and $r_{21} = 0.4$, calculate the value of $r_{12.3}$.

Solution:

$$r_{12.3} = \frac{r_{12} - r_{13}\, r_{23}}{\sqrt{1 - r^2_{13}}\sqrt{1 - r^2_{23}}}$$

$r_{12} = 0.65,\ r_{23} = 0.6,\ r_{13} = 0.4$

Substituting the values

$$r_{12.3} = \frac{0.4-.65\times.6}{\sqrt{1-(.65)^2}\sqrt{1-(.6)^2}}$$

$$= \frac{.4-.39}{\sqrt{1-.4225}\sqrt{1-.36}} = \frac{0.01}{.8\times.64} = \frac{0.01}{0.512} = 0.02.$$

Example 18:

The table shows the corresponding values of three variables, X_1, X_2 and X_3. Find the least square regression equation of X_3 on X_1 and X_2. Estimate X_2 when X_1 = 10 and X_2 = 6.

$\overline{X}_1$	3	5	6	8	12	14
$\overline{X}_2$	16	10	7	4	3	2
$\overline{X}_3$	90	72	54	42	30	12

Solution:

The regression equation of X3 on X2 and X1 can be written as follows:

$$X_3 - \overline{X}_3 = \left(\frac{r_{23} - r_{13}\, r_{12}}{1 - r^2_{12}}\right)\left(\frac{S_3}{S_1}\right)(X_2 - \overline{X}_2)$$

$$+ \left(\frac{r_{13} - r_{23}\, r_{12}}{1 - r^2_{13}}\right)\left(\frac{S_3}{S_1}\right)(X_1 - \overline{X}_1)$$

Calculating $\overline{X}_1$, $\overline{X}_2$, $\overline{X}_3$, S_1, S_2, S_3, r_{12}, r_{13}, r_{23}

X_1	$(X_1-\overline{X}_1)$	X_1^2	X_2	$(X_2-\overline{X}_2)$ x^2	X_2^2	X_3	$(X_3-\overline{X}_3)$ x_3	X_3^2	X_1X_2	X_1X_3	X_2X_3
3	−5	25	16	+9	81	90	+40	1600	−45	−200	+360
5	−3	9	10	+3	9	72	+22	484	−9	−66	+66
6	−2	4	7	0	0	54	+4	16	0	−8	0
8	0	0	4	−3	9	42	−8	64	0	0	+24
12	+4	16	3	−4	16	30	−20	400	−16	−80	+80
14	+6	36	2	−5	25	12	−38	1444	−30	−228	+190
ΣX_1 =48	ΣX_1 =0	ΣX_1^2 =90	ΣX_2 =42	ΣX_2 =0	ΣX_2^2 =140	ΣX_3 =300	ΣX_3 =0	ΣX_3^2 =4008	ΣX_1X_2 =−100	ΣX_1X_3 =−582	ΣX_2X_3 =720

$$\overline{X}_1 = \frac{48}{6} = 8,\ \overline{X}_2 = \frac{42}{6} = 7,\ \overline{X}_3 = \frac{300}{6} = 50$$

$$S_1 = \frac{\sqrt{\sum\left(X_1 - \bar{X}_1\right)^2}}{N} = \frac{\sqrt{90}}{6} = \sqrt{15} = 3.87$$

$$S_2 = \frac{\sqrt{\sum\left(X_2 - \bar{X}_2\right)^2}}{N} = \frac{\sqrt{140}}{6} = \sqrt{23.33} = 4.83$$

$$S_3 = \frac{\sqrt{\sum\left(X_3 - \bar{X}_3\right)^2}}{N} = \frac{\sqrt{4008}}{6} = \sqrt{668} = 25.85$$

$$r_{12} = \frac{\sum X_1 X_2}{\sqrt{\sum X_1^2 \times \sum X_2^2}} = \frac{-100}{\sqrt{90 \times 140}} = -0.891$$

$$r_{13} = \frac{\sum X_1 X_3}{\sqrt{\sum X_1^2 \times \sum X_3^2}} = \frac{-582}{\sqrt{90 \times 4008}} = 0.969$$

$$r_{23} = \frac{\sum X_2 X_3}{\sqrt{\sum X_2^2 \times \sum X_3^2}} = \frac{720}{\sqrt{140 \times 4008}} = 0.961$$

$$X_3 - 50 = \left[\frac{0.961 - (-0.969 \times -0.891)}{1 - (-0.9)^2}\right]\left(\frac{25.85}{4.83}\right)(X_2 - 7)$$

$$+ \left[\frac{0.969 - (-0.961 \times -0.891)}{1 - (-0.9)^2}\right]\left(\frac{25.85}{3.87}\right)(x_1 - 8)$$

$$x_3 - 50 = 2.546\ (x_2 - 7) - 3.664\ (x_1 - 8)$$

$$x_3 - 50 = 2.546\ x_2 - 17.822 - 3.664\ x_1 + 29.312$$

$$x_3 = 2.546\ x_2 - 3.664\ x_1 + 61.49$$

$x_1 = 10$ and $x_2 = 6$, X_3 will be:

when $x_3 = 15.276 - 36.64 + 61.49 = 40.126$ or 40.

Example 19:

In a trivariate distribution, it is found that $r_{13} = 0.6$, $r_{23} = 0.5$, $r_{12} = 0.8$. *Find the value of* $r_{13.2}$.

Solution:

$$r_{13.2} = \frac{r_{13} - r_{12}\, r_{23}}{\sqrt{1 - r^2_{13}}\sqrt{1 - r^2_{23}}}$$

$$r_{13} = .6,\ r_{12} = .8,\ r_{23} = .5$$

$$r_{13.2} = \frac{.6-(.8)(.5)}{\sqrt{1-(.8)^2}\sqrt{1-(.5)^2}}$$

$$= \frac{.6-.4}{\sqrt{.36}\sqrt{.75}} = \frac{.2}{.5196} = 0.385$$

Example 20:

Is it possible to get the following fro

m a set of experimental data?

$r_{12} = 0.6,\ r_{23} = 0.8,\ r_{31} = -0.5$

Solution:

$$r_{12.3} = \frac{r_{12} - r_{13}\, r_{23}}{\sqrt{1-r^2{}_{13}}\sqrt{1-r^2{}_{23}}}$$

$$= \frac{.6-(-.5)\times(.8)}{\sqrt{1-(.5)^2}\sqrt{1-(.8)^2}}$$

$$= \frac{.6+.4}{\sqrt{.75\times.36}} = \frac{1}{.52} = 1.92$$

The value of $r_{12.3}$ cannot be greater than one. Since, in the given case $r_{12.3}$ has exceeded one, there is some inconsistency in the given data.

Example 21:

If $r_{12} = 0.60,\ r_{13} = 0.70,\ r_{23} = 0.65$ *and* $S_1 = 1.0$, *find* $S_{1.23}$, $R_{1.23}$ *and* $r_{12.3}$.

Solution:

$$S_{1.23} = S_1 \frac{\sqrt{1-r^2{}_{12} - r^2{}_{13} - r^2{}_{23} + 2r_{12}\, r_{13}\, r_{23}}}{1-r^2{}_{23}}$$

$$= 1.0 \frac{\sqrt{1-(.6)^2-(.7)^2-(.65)^2+2\times.6\times.7\times.65}}{1-(.65)^2}$$

$$= 1.0 \frac{\sqrt{1-0.36-0.49-.423+0.546}}{0.577}$$

$$= 1.0\frac{\sqrt{0.273}}{0.577} = 0.688$$

$$R_{1.23} = \frac{\sqrt{r^2{}_{12} + r^2{}_{13} - 2\, r_{12}\, r_{13}\, r_{23}}}{1 - r^2{}_{23}}$$

$$= \frac{\sqrt{(.6)^2 + (.7)^2 - 2(.6)(.7)(.65)}}{1 - (.65)^2} = \frac{\sqrt{.304}}{.5775} = 0.726$$

$$r_{12.3} = \frac{r_{12} - r_{13}\, r_{23}}{\sqrt{1 - r^2{}_{13}}\sqrt{1 - r^2{}_{23}}}$$

$$= \frac{.6 - .7 \times .65}{\sqrt{1 - (.7)^2}\sqrt{1 - (.65)^2}}$$

$$= \frac{.6 - .455}{\sqrt{.51 \times 5775}} = \frac{.145}{.543} = 0.267.$$

Example 22:

The simple correlation coefficients between profits (X_1), sales (X_2) and advertising expenditure (X_3) of a factory are r_{12} = 0.69, r_{13} = 0.45 and r_{23} = 0.58. Find the partial correlation coefficients $r_{12.3}$ and $r_{13.3}$ and interpret them.

Solution:

$$R_{12.3} = \frac{r_{12} - r_{13}\, r_{23}}{\sqrt{1 - r^2{}_{12}}\sqrt{1 - r^2{}_{23}}}$$

$$= \frac{.69 - .45 \times .58}{\sqrt{1 - (.45)^2}\sqrt{1 - (.58)^2}}$$

$$= \frac{.69 - .261}{\sqrt{1 - 2022}\sqrt{1 - .3364}}$$

$$= \frac{.429}{\sqrt{.7975 \times 6636}} = \frac{.429}{.7275} = 0.598$$

$$r_{13.2} = \frac{r_{13} - r_{12}\, r_{23}}{\sqrt{1 - r^2{}_{12}}\sqrt{1 - r^2{}_{23}}}$$

$$= \frac{.45 - .69 \times .58}{\sqrt{1-(.69)^2}\sqrt{1-(.58)^2}}$$

$$= \frac{.45 - .4002}{\sqrt{1-.4761}\sqrt{1-.3364}} = \frac{.0498}{\sqrt{.5239 \times .6636}} = 0.085$$

Example 23:

If $r_{12} = 0.80$, $r_{13} = -0.56$ *and* $r_{23} = 0.40$, *then obtain* $r_{12.3}$ *and* $R_{1.23}$.

Solution:

$$r_{12.3} = \frac{r_{12} - r_{13}\, r_{23}}{\sqrt{1 - r^2_{13}}\sqrt{1 - r^2_{23}}}$$

$$r_{12} = 0.8,\ r_{13} = -0.56,\ r_{23} = -0.4$$

$$r_{12.3} = \frac{.8 - (-.56)(-.4)}{\sqrt{1-(-.56)^2}\sqrt{1-(-.4)^2}}$$

$$= \frac{.8 - .224}{\sqrt{.6864}\sqrt{.84}} = \frac{.576}{.759} = 0.759$$

$$R_{1.23} = \frac{\sqrt{r^2_{12} + r^2_{13} - 2\, r_{12}\, r_{13}\, r_{23}}}{1 - r^2_{23}}$$

$$= \frac{\sqrt{(.8)^2 + (-.56)^2 - 2(.8)(-.56)(-.4)}}{1-(-.4)^2}$$

$$= \frac{\sqrt{.64 + .3136 - .3584}}{1 - .16} = \frac{\sqrt{.5952}}{.84} = 0.842$$

Example 24:

If $r_{12} = 0.5$, $r_{31} = 0.3$, $r_{23} = 0.45$, *find* $R_{3.12}$.

Solution:

$$R_{3.12} = \frac{\sqrt{r^2_{13} + r^2_{23} - 2\, r_{12}\, r_{13}\, r_{23}}}{1 - r^2_{13}}$$

$$= \frac{\sqrt{(0.3)^2 + (0.45)^2 - 2(0.5 \times 0.3 \times 0.45)}}{1-(0.5)^2}$$

$$= \frac{\sqrt{0.09 + 0.2025 - 2(0.0675)}}{1 - 0.25}$$

$$= \sqrt{0.57} = 0.755.$$

Example 25:

If $r_{12} = 0.77$, $r_{13} = 0.72$ *and* $r_{23} = 0.52$, *find the partial correlation coefficient* $r_{12.3}$ *and multiple correlation coefficient* $R_{1.23}$.

Solution:

$$r_{12.3} = \frac{r_{12} - r_{13}\, r_{23}}{\sqrt{1 - r^2_{13}}\,\sqrt{1 - r^2_{23}}}$$

$$r_{12} = 0.77,$$

$$r_{13} = 0.72,$$

$$r_{23} = 0.52$$

$$r_{12.3} = \frac{.77 - .72 \times .52}{\sqrt{1 - (.72)^2}\,\sqrt{1 - (0.52)^2}}$$

$$= \frac{.77 - .37}{\sqrt{1 - .5184}\,\sqrt{1 - .2704}}$$

$$= \frac{.4}{\sqrt{.4816 \times .7296}} = \frac{.4}{.593} = 0.674$$

$$R_{1.23} = \frac{\sqrt{r^2_{12} + r^2_{13} - 2\, r_{12}\, r_{13}\, r_{23}}}{1 - r^2_{23}}$$

$$= \frac{\sqrt{(.77)^2 + (.72)^2 - 2(.77)(.72)(.52)}}{1 - (.52)^2}$$

$$= \frac{\sqrt{.5929 + .5184 - .5766}}{1 - .2704}$$

$$= \frac{\sqrt{.5347}}{.7296} = 0.856.$$

LIST OF FORMULAES

Partial Correlation Coefficients $r_{12.3} = \dfrac{r_{12} - r_{13}\, r_{23}}{\sqrt{1 - r_{13}^2}\ \sqrt{1 - r_{23}^2}}$

$$r_{13.2} = \frac{r_{13} - r_{12}\, r_{23}}{\sqrt{1 - r_{12}^2}\ \sqrt{1 - r_{23}^2}}$$

$$r_{23.1} = \frac{r_{23} - r_{12}\, r_{13}}{\sqrt{1 - r_{12}^2}\ \sqrt{1 - r_{13}^2}}$$

Multiple Correlation Coefficient $R_{1.23} = \dfrac{\sqrt{r_{12}^2 - r_{13}^2 - 2r_{13}\, r_{12}\, r_{23}}}{1 - r_{23}^2}$

Multiple Regression of X_1 on X_2 and X_3 $X_{1.23} = a^{1.23} + b_{12.3}\, X^2 + b_{13.2}\, X^3$ $b_{12.3} = \dfrac{\sigma_1}{\sigma_2} \times \dfrac{r_{12} - r_{13}\, r_{23}}{1 - r^2{}_{23}}$

Partial Regression Coefficient $b_{13.2} = \dfrac{\sigma_1}{\sigma_2} \times \dfrac{r_{13} - r_{12}\, r_{23}}{1 - r^2{}_{23}}$

EXERCISES

1. The table shows the corresponding values of three variables X_1, X_2 and X_3 :

X_1 :	3	5	6	8	12	14
X_2 :	16	10	7	4	3	2
X_3 :	90	72	54	42	30	12

   Find the least squares regression equation of X_3 on X_1 and X_2. Estimate X_3 when $X_1 = 10$ and $X_2 = 6$.

2. $b_{12.3}$ is called the partial regression coefficient of :
   (i) X_2 on X_3, keeping X_1 constant.
   (ii) X_1 on X_3, keeping X_3 constant.
   (iii) X_1 on X_3, keeping X_2 constant.
   (iv) none of these.

3. Find the multiple linear regression of X_1 on X_2 and X_3 from the data relating to three variables given below :

X_1 :	11	17	26	28	31	35	41	49	63	69
X_2 :	2	4	6	5	8	7	10	11	13	14
X_3 :	2	3	4	5	6	7	8	10	11	13

4. Estimate the relationship between the use of inputs and labour on productivity from the following data :

Productivity :	15	18	16	20	24	27
Input :	5	8	7	6	10	9
Labour :	40	45	50	55	60	50

   (i) What will be the productivity if inputs will be 12 and labour 65?

   (ii) Compute the coefficient of multiple determination.

5. (a) Distinguish between simple, multiple and partial correlations.

   (b) What is multiple linear regression? What is its importance in economic analysis?

   (c) When is multiple regression needed? Explain with the help of an example.

6. (a) What is meant by multi-collinearity?

   (b) Explain the concept of multiple regression and cite an example of its utility in the practical field.

7. Tick the correct answer :

   (a) In multiple correlation analysis, there are at least

   (i) 2 variables,

   (ii) 3 variables,

   (iii) 5 variables,

   (iv) 10 variables,

   (v) none of these.

   (b) $r_{13.2}$ means are coefficient of partial correlation between

   (i) X_1 and X_2, keeping the effect of X_3 constant,

   (ii) X_2 and X_3, keeping the effect of X_1 constant,

   (iii) X_1 and X_2, keeping the effect of X_2 constant,

   (iv) X_1, X_2 and X_3.

(c) In case of 3 variables, X_1, X_2 and X_3, there are

(i) 3 partial correlation coefficients,

(ii) 2 partial correlation coefficients,

(iii) 4 partial correlation coefficients,

(iv) at least 10.

(d) $r_{12.4}$ is calculated as :

(i) $\dfrac{r_{12} - r_{13}\, r_{24}}{\sqrt{\left(1 + r_{13}^2\right)\left(1 - r_{24}^2\right)}}$ (ii) $\dfrac{r_{13} - r_{12}\, r_{24}}{\sqrt{\left(1 + r_{12}^2\right)\left(1 - r_{24}^2\right)}}$

(iii) $\dfrac{r_{12} + r_{13}\, r_{24}}{\sqrt{\left(1 + r_{13}^2\right)\left(1 - r_{24}^2\right)}}$ (iv) $\dfrac{r_{24} - r_{21}\, r_{14}}{\sqrt{\left(1 - r_{12}^2\right)\left(1 - r_{11}^2\right)}}$

(v) $\dfrac{r_{13} - r_{14}\, r_{24}}{\sqrt{\left(1 - r_{14}^2\right)\left(1 - r_{24}^2\right)}}$

8. Fill in the blanks :

(a) Partial correlation coefficients such as $r_{12.3}$ and $r_{13.2}$ are referred to as ....

(b) Second order partial correlation coefficients can be obtained from .... coefficients.

(c) By squaring $R_{1.23}$ we obtain the coefficient of ....

(d) The coefficient of multiple correlation lies between .... and ....

(e) The dependent variable is always denoted by ....

(a) First order coefficients,

(b) first order,

(c) multiple determination,

(d) 0, 1, (e) X_1.

9. (a) The following information about a trivariate population is known:

$\sigma_1 = 3\ \sigma_2 = 4$ and $\sigma_3 = 5$

$r_{23} = 0.40\ r_{31} = 0.60$ and $r_{12} = 0.7$

Determine the regression equation of X_1 on X_2 and X_3.

(b) If $r_{12} = 0.9$, $r_{13} = 0.75$, $r_{23} = 0.7$, find $R_{1.23}$.

10. Find the multiple regression equation of X_1 and X_3 from the data relating to three variables given below :

X_1	X_2	X_3	X_1	X_2	X_3
4	15	30	9	6	14
6	12	24	13	4	10
7	8	20	15	3	4

Also predict the value of X_1 when $X_2 = 10$ and $X_3 = 22$.

11. The coefficient of multiple correlation between X_2 on the one hand and X_1 and X_3 on the other is denoted by

(i) $r_{12.3}$ (ii) $r_{13.2}$ (iii) $R_{1.23}$ (iv) $R_{2.13}$ (v) $R_{3.12}$

12. In case of three variables, the regression equation X_1 on X_2 and X_3 has the form :

(i) $X_1 = a_{1.23} + b_{12.3}\ X_2 + b_{13.2}\ X_3$

(ii) $X_1 = a_{2.13} - b_{12.3}X_2 + b_{12.3}X_3$

(iii) $X_1 = a_{1.22} + b_{12.3}X_2 + b_{13.2}X_3$

(iv) $X_2 = a_{1.23} + b_{12.3}X_2 - b_{13.2}X_1$

(v) none of these.